Basics of
ENVIRONMENTAL SCIENCE
and
ENGINEERING

NIPA® GENX ELECTRONIC RESOURCES & SOLUTIONS P. LTD.

New Delhi-110 034

Basics of Environmental Science *and* Engineering

P. SIVASHANMUGAM
Assistant Professor
Department of Chemical Engineering
National Institute of Technology
Tiruchirapalli - 620 015
Tamil Nadu (India)

NIPA® GENX ELECTRONIC RESOURCES & SOLUTIONS P. LTD.
New Delhi-110 034

NIPA® GENX ELECTRONIC RESOURCES & SOLUTIONS P. LTD.

101,103, Vikas Surya Plaza, CU Block
L.S.C. Market, Pitam Pura, New Delhi-110 034
Ph : +91 11 27341616, 27341717, 27341718
E-mail: newindiapublishingagency@gmail.com
www: www.nipabooks.com

For customer assistance, please contact
Phone: + 91-11-27 34 17 17
Fax: + 91-11- 27 34 16 16
E-Mail: feedbacks@nipabooks.com

ISBN: 978-81-19002-79-5

Composed and Designed by NIPA®.

Preface

Environmental science is the interdisciplinary field and requires the study of the interactions among the physical, chemical and biological components of the Environment; with a focus on Environmental pollution and degradation of the Environment related due to human activities; and the impact on Biodiversity and sustainability from local and global development. It is inherently an interdisciplinary field that draws upon not only its core scientific areas, but also applies knowledge from other non-scientific studies such as economics, law and social sciences.

Even though the concept of environmental science has existed for centuries, it become more active during the decades of 1960s and 1970s, for the analysis of complex Environmental problems with substantive Environmental laws and the support of the growing public awareness on Environmental problems.

The population and increase in standard of living is also one of the major factor which makes the changes on Environmental issues such as Climate change, Conservation, Biodiversity, Groundwater and Soil contamination, Use of natural resources, Waste Management, Sustainable development, and Air pollution.

The book will provide complete overview of the status and role of various resources on Environment. The book is prepared with simple approach on various factors for Undergraduate/Postgraduate students level. This book will be useful for Engineering as well as science graduates for their curriculum. All efforts are made to cover the various topics on Environment issues with some suitable examples. The book can also be used for any level of competitive examination.

The book divided in to five units and the *Unit I* for the Introduction to Environmental Studies and Natural resources, *Unit II* for the Ecosystems and Biodiversity, *Unit III* for Environmental Pollution, *Unit IV* for Social Issues and the Environment and *Unit V* for Human population and the Environment.

Care has been taken to **cover the syllabus of Indian universities as per UGC and AICTE** norms so that the students can make use of this book as text and for the examination point of view.

P. Sivashanmugam

Contents

CHAPTER 1

Introduction to Environmental Studies and Natural Resources

1.1 Forest Resources

Forest resources usually provide the raw materials for homes, workplaces, the books and newspapers, and the packaging that contains food and other products of our labor. Forest ecosystems supply our water, maintain our climate, and help purify the air, protect soils, and provide for wilderness experiences. Forests provide habitat for wildlife, and serve as preserves of biological diversity. Forests are sources of food, fuel and medicine for people all over the world. Forests shape the recreational landscape, help stabilize our farms, and enhance our cities.

About 30 per cent of the world's total land areas are covered by forest with the area of 3.9 billion hectares. Nearly 47 percent of the world's forests are in the tropics, 33 percent in the boreal zone, 11 percent in the temperate zones, and 9 percent in sub-tropical areas. The 94 million hectares of forest lost over the ten-year period, represented about 2 per cent of the world's total forest cover, or an area larger than Venezuela. Two thirds of the world's forests are located in ten countries: the Russian Federation, Brazil, Canada, the United States, China, Australia, the Democratic Republic of the Congo, Indonesia, Angola and Peru. Most of the deforestation work was performed in natural tropical forests, which lost 14.2 million hectares a year over the last decade. Africa and South America have suffered the most deforestation.

Africa was the region with the highest deforestation in the world, which lost 5.3 million hectares of forest per year in the 1990s. Forests are a major factor in the climate change issue. Forest ecosystems contain more than half of all terrestrial carbon, and account for about 80 per cent of the exchange of carbon between terrestrial ecosystems and the atmosphere. Deforestation in the 1980s may have accounted for a quarter of all human-induced carbon emissions, the second greatest emitter after fossil fuels. Forest plantations comprise 5 percent of the world's forests. Asia has the largest area of plantations, accounting for 62 per cent of the world total. China accounts for 24 percent of that total and India, 18 per cent. Plantations supply about 35 percent of the world's round wood. The area of forest plantations increased by an average of 3 to 4 million hectares per year during the 1990s. Half of this increase was the result of afforestation on land previously under non-forest land use, whereas the other half resulted from conversion of natural forest. There are reports that 12 percent of the world's forests, or about 480 million hectares, are in protected areas.

1.2 Uses of Forest

Many forest trees species that provide goods and services such as firewood, fruits, timber, poles, fodder, environmental protection, amenities, Let us see some uses of forest.

(i) Trees for Firewood

Although, practically all tree species could be used for firewood, there is a group of species that is generally preferred. Such species usually burn without excessive smoke and unpleasant odors. The choice of species is however now very restricted in some localities due to unavailability of some of the preferred species in sufficient quantities. Wood is also used in the commercial production of charcoal particularly in South Africa, India, China and Zimbabwe.

(ii) Trees used for Food Production

Forest can be sources of food from their fruits, more so in times of food shortage. Forest tree species also directly provide other edible products such as seed and nuts, gums and resins, root tubers and leaves for vegetables. Indigenous fruits, for example are processed into a variety of products that store well and therefore readily available in periods of food shortage. Some species are used for the production of alcoholic beverages. The Amarula cream, an alcoholic product developed from *S. birrea* fruit in South Africa is now marketed in about 50 countries.

(iii) Non-wood Forest Products from Trees and Forests

Non-wood forest products such as edible worms and insects and mushrooms can be obtained from variety of species from the forest. In Zambia, 38 indigenous tree species are known to produce tannins, 19 produce dyes and 11 species produce resins and gums. In 1992, Zambia produced honey and beeswax amounting to 90 tonnes and 29 tonnes respectively which was valued at US$170000 and US$74000.

(iv) Trees used for Fodder

A wide range of tree species is regarded as important fodder tree species. The indigenous tree species cited as being widely used for fodder.

A lot of research is focussed to develop high yield fodder for the growing demand of milk.

(v) Trees used for Poles

Many indigenous and exotic species are mainly used in building huts, cattle pens, fencing and general farm construction. The most widely used species are *Androstachys johnsonii, B. plurijuga, Terminalia sericea*. The most durable and termite resistant species are the most preferred. Among the exotic species, the mostly widely used species for making poles are *C. cunninghamiana, E. camaldulensis, E. citriodora.*

(vi) Trees for Shade and Shelter

The indigenous species *A. erioloba, A. quanzensis, Albizia versicolor, Celtis africana* are being used for giving shade The exotic tree species used for shade and shelter are *Toona ciliata, C. cunninghamiana, Eucalyptus* spp., *J. mimosifolia, Pinus* spp. and *Thuya orientalis.*

(vii) Trees used for Timber

Most of the countries in the region do not have naturally occurring fast growing tree species that could be exploited to meet the region's growing demand for timber, pulp and tannins. Whereas, Malawi has small areas of exploitable conifers such as *W. nodiflora*, which is widely used for timber production. The commercially exploited indigenous timber species include *A. quanzensis, Allanblackia stuhlmanii.*

(viii) Trees used for Medicinal Purposes

There are many tree species used in the treatment of a range of ailments. These species vary from country to country. The most commonly used species are *Cassipourea malosana, Cassia abbreviata, Dioscorea sylvatica, Diplorrynchus* and The value of medicinal plants in Namibia is estimated at US$3.2 million. In Mauritius, over 90 indigenous trees and shrubs are used for medicinal purposes. The traditional medicine industry in South Africa is estimated at between US$50 million and $100 million annually.

(ix) Trees used for Amenities and Aesthetic Purposes

Most of these amenity tree species are either planted in gardens, along avenues in urban centers or as hedges.

(x) Trees used for conservation/environmental protection

A wide variety of forest trees are used for environmental protection such as gully reclamation, restoration of degraded sites, fixing shifting dunes, etc.

1.3 Deforestation and Its Effects

Deforestation is the removal or damage of vegetation in a forest such that it no longer supports its natural flora and fauna. This is quite often done by humans taking care of immediate needs without knowing the long-term effects of their actions. Deforestation is a slow-onset disaster that may contribute to other, cataclysmic disasters. Catastrophic proportions of disasters may be expected after large areas of vegetation in a forest are damaged or removed. Some times land's protective and regenerative properties may get spoiled due to change in area's of natural flora and fauna.

According to a Food and Agriculture Organization and United Nations Environmental Program (FAO/UNEP) study in 2001, tropical forests are disappearing at the rate of 7.3 million hectares per year. It is estimated that about 4.2 million hectares ie 10.4 million acres a year in Latin America, 1.8 million hectares ie 4.4 million acres a year in Asia and 1.3 million hectares ie 3.2 million acres a year in Africa.

Case Studies

(i) Deforestation in Africa

There is a report that 2.3 million hectares of open woodlands are cleared each year in Africa and the vegetation of other vast areas of woodlands is declining with severe consequences for land and people. More over, the survival of extensive forests in one region provides little comfort to residents of areas experiencing wholesale forest destruction. The dense forests of West Africa and a few other areas in the continent (especially Madagascar, Rwanda, and Burundi) are disappearing fast, while rain forests in parts of Central Africa are little touched as yet. The Ivory Coast has lost 70 percent of the forest with which it began the twentieth century; there and in Nigeria some 10 percent of the accessible forest was cleared each year in the second half of the 1970s.

(ii) Deforestation in Asia

Recent satellite pictures of the Philippines, traditionally a major timber exporter, indicate that forests now cover only 30 percent of the country. Actually the Philippines government feels a forest cover of 46 percent is desirable for economic and environmental reasons. If existing logging patterns continues, all original old- growth forests will have been cut down by the year 2010, and projected timber supplies from second-growth forests and plantations will not suffice to meet even domestic needs. Destructive increases in flooding and sedimentation have already been registered. Indonesia, which emerged in the 1970s as the world's leading tropical-timber exporter, retains extensive and rich forests on its outer islands. Now the Malaysia is experiencing this problem.

1.3.1 Effects of Deforestation

Deforestation causes many effects, out of which biodiversity and climate change is being regarded as major effects, which are described as follows.

(i) Biodivesity

The process of deforestation in various geographical locations is deteriorating the unique environment. Consequently, many animals and plants that live in the rainforests face the specter of extinction. The extinction of the plants and animals leads to diminished gene pool. The lack of biodiversity and a reduced planetary gene pool may have many unforeseen effects, some of which could be fatal to the future of humanity. Besides, there are ethical, aesthetic and philosophical question regarding mankind's responsibility for other life.

(ii) Climate Change

When an area of rainforest is either cut down or destroyed, various climate changes have been reported due to deforestation.. The following is a list of the various climate changes due to deforestation.

a. *Desiccation of previously moist forest soil :* Due to exposure to the sun, the soil in the forest area gets baked and the lack of canopy leaves nothing to prevent the moisture from quickly evaporating into the atmosphere. So, previously moist soil becomes dry and cracked.

b. *Dramatic Increase in Temperature Extremes :* Trees provide shade and the shaded area has a moderated temperature. With shade, the temperature may be 103 degrees Fahrenheit during the day and 60 degrees at night. With out the shade, temperatures would be much colder during the night and around 130 degrees during the day.

c. *Moist Humid Region Changes to Desert :* Mainly because of the lack of moisture and the inability to keep moisture, soil exposed to the sun will dry and turn into desert sand. Even before that happens, when the soil becomes dry, dust storms become more frequent. At this stage, the soil becomes useless.

d. *No Recycling of Water :* Moisture from the oceans fall as rain on adjacent coastal regions. The moisture is soon sent up to the atmosphere through the transpiration of foliage to fall again on inland forest areas. This cycle repeats several times to rain on all forest regions.

e. *Less Carbon Dioxide and Nitrogen Exchange :* The rainforests are important in the carbon dioxide exchange process. They are second only to oceans as the most important "sink" for atmospheric carbon dioxide. There are reports that deforestation may account for as much as 10% of current greenhouse gas emissions. Greenhouse gases are gases in the atmosphere that literally trap heat. There is a theory that as more greenhouse gasses are released into the atmosphere, more heat gets trapped. Thus, there is a global warming trend in which the average temperature becomes progressively higher.

f. *More Desertification :* Deforestation is an important factor contributing to desertification. According to a report between 1958 and 1975, the Saharan Desert expanded southward by about 100km. It is estimated that desertification threatened 35 per cent of the world's land surface and 20 per cent of the world's population.

g. *Soil Erosion :* Deforestation is known to contribute to run-off of rainfall and intensified soil erosion. The seriousness of the problem depends much on soil characteristics and topography.

h. *Other Effects :* Forest provides two most important things for the living creature is clean air and clean water. Rainforests also provide many aesthetic, recreational and cultural rewards. If these rainforests are destroyed, then these rewards disappear. This has major social repercussions for the entire world.

1.4 Water Resources

Water resource is one of the most important among other natural resources. Water is the widely distributed substance on our planet; in different amounts it is available everywhere and plays an important role in the surrounding environment and human life. Water exists in three states namely liquid, solid, and vapor and forms oceans, seas, lakes, rivers, and ground waters in the top layers of Earth's crust and soil cover. Water is available in solid form namely ice and snow in polar region and in vapor, water droplet form, ice crystal form in biosphere. Water states formations is very dynamic in the globe making it difficult to get the proper estimate of the resources. The water being in state of liquid, solid or gaseous state in the atmosphere, on the Earth's surface and in crust down to the depth of 2000 meters called hydrosphere.

It is reported that the Earth's hydrosphere contains a huge amount of water of about 1386 million cubic kilometers, of which 97.5% is saline water and only 2.5% is fresh water. The greater portion of the fresh water (68.7%) is in the shape of ice and permanent snow cover in the Antarctic, the Arctic, and mountainous regions. Next 29.9% are fresh ground waters. Only 0.26% of the total amount of fresh waters on the Earth are concentrated in lakes, reservoirs, and river systems.

Solar heat evaporates water into the air from the Earth's surface. Land, lakes, streams, and oceans send up a steady stream of water vapor. The evaporated vapor spreads over the surface of our planet falling down as precipitation. Precipitation falling down on land is the main source of the formation of land waters: rivers, lakes, groundwater, and glaciers. Also some portion of the precipitated water penetrates and charge groundwater, and the other portion flow as river and returns to the ocean where again evaporates, and the process repeats again and again. A considerable portion of river flow does not reach the ocean having evaporated in the so-called drainless regions, or the regions of endorheic runoff. On the other hand, some part of groundwater passing by the river systems, directly goes to the ocean or evaporates.

Every year the water turnover on Earth involves 577,000 km^3 of water. It is the water that evaporates from the oceanic surface (502,800 km^3) and from land (74,200 km^3). The same water amount falls as atmospheric precipitation (on the ocean 458,000 km3 and on land 119,000 km^3). The difference between precipitation and evaporation from land surface (119,000 - 74,200 = 44,800 km^3/year) represents the total runoff of Earth's rivers (42,600 km3/year), and a direct groundwater runoff to the ocean (2200 km^3/year).

Hydrospheric water of different types are fully replenished during different period in the process of hydrological cycle as World Ocean- 2500 years ,Ground water- 1400 years ,Polar ice-9700 years , Mountain glaciers-1600 years, Ground ice of the permafrost zone-10000, Lakes-17 years, Bogs-5 years , Soil moisture-1 year, Channel network-16 days , Atmospheric moisture- 8 days , Biological water-several hours,

The renewable water resources means the water yearly replenished in the process of water turnover on the earth. It is mainly the river runoff estimated in the volume referred to a unit of time (m^3/s, km^3/year, etc.) and formed in the region at issue or incoming from outside, including the groundwater inflow to the river network.

The water resource in a particular region are influenced by many factors as follows

(i) Climatic factors: Under this rainfall intensity, duration & distribution, Snow fall intensity and duration & Evapo-transpiration will be major factors

(ii) Physiographic factors: In this (a) Geometric factors such as drainage area, shape, slope & stream density, (b) Physical factors such as *land* use, surface infiltration conditions, *soil* types will be major factors

(iii) The following geological factors also affect the water resources in a particular region

(a) Lithologic including composition, texture, sequence of rock types and the thickness of rock formations.

(b) Structural, including chief faults & folds that interrupt the uniformity of occurence of rock types or sequence of rock types also beds, joints, fissures, cracks, etc.

(c) Hydrologic characteristics of the aquifers permeability, porosity, transmissivity, storability, etc

The physiographic and geological factors influence the occurrence and distribution of water resources within a region. Orography also play a significant role in influencing rainfall, temperature, humidity and wind. Also within a geographical location and physiographic framework, the rainfall intensity, duration and distribution and the climatic factors affecting evapo-transpiration that determine the totality of water resources in the particular region.

1.4.1 Water Conflicts

The relating of water availability in different parts of a basin, and the linkage between land use and water use in one river basin can have profound effects on other activities in different parts of the basin. Proper understanding of this interconnection is needed to cope up water problem. The failure to perceive this interconnectedness in the planning of water utilization will lead to a major source of conflict over water use in river basins.

Water allocation for demands usually will be done without knowing the importance of the demand leading to aggravation of the problems of inequalities and maldistribution. Water planning is done on four major sector categories of use namely (i) Domestic (ii) Agricultural (iii) Industrial (iv) Power generation.

Internal conflicts

Conflicts can be internal within the sector or between the sectors. Usually multi-purpose river valley projects are primarily aimed for both irrigation and power generation. Generally problem will not arise as far as purpose is concerned. However, location of these projects will play a major in determining these objectives and water releases are also determined by priorities for power or for irrigation. Conflicts between the sector arise due to diversion of water for irrigation from drinking water, or for industry from agricultural and domestic use. Conflicts also arise on the basis of conflicting interests between the rich and powerful and the poor and marginal. The category 'domestic' as an undifferentiated one conceals the conflict between the poor rural peasants requiring a pot full of drinking water and the rich urban elite using large quantities of water for meeting the requirements of water-intensive sewage systems, space cooling, gardening, etc. Domestic requirements vary for different people and the high demands from urban areas are often met by diverting water from rural areas. Conflicts also arise due to use of water for types agriculture. For example there is a conflicts between water-intensive cultivation of commercial crops for high cash returns and prudent water use for protective irrigation of staple food crops essential for survival.

Social conflicts

Social conflicts over water can also be analyzed at different societal levels. For example the inter-state conflicts are generated when water projects of upstream states influence the quality and quantity of water flow in the basin and reduce the possibilities of water use by downstream states. Major inter-basin water transfers in rivers flowing through many states also generate conflicts by disturbing the riparian rights of states. Conflicts also arise between the state and the people when official planning and policies lead to changes in water use and utilisation pattern and therefore undermine people's access to water. This is being experienced in states between Tamil Nadu and Karnataka, Tamil Nadu and Kerala, Andra Paradesh and Karnataka, Punjab and Haryana. Also state planned quarrying of minerals or timber extraction in the river catchments affect the river flow and generate conflicts downstream. Similarly, state planned agricultural production based on large irrigation projects to generate marketable surpluses of cash crops conflicts with people's needs for local food production. Such projects also lead to conflicts between the state and the people by eroding traditional water rights which are often communal in nature and ensure the survival of all members of the community. As far as surface and groundwater use is concerned, state intervention has led to the concentration of water access in the hands of the rich thereby generating new conflicts between the rich and poor.

Case study

The Kabini project has a submersion area of nearly 6,600 acres, but it led to the clear felling of 30,000 to 35

0000 acres of primeval forests in the catchments to rehabilitate displaced villages. Due to this, local annual rainfall fell from 65-60 inches to 45 -40 inches, and high siltation rates have drastically reduced the life of the project. In this project upstream side of Cauvery river, large areas of well developed coconut gardens and paddy fields have been laid waste through water-logging and salinity within two years of irrigation from the project. The Kabini project is a classic case of how the water crisis is created by the very projects aimed at increasing water availability or stabilizing water flows in the engineering paradigm.

1.4.2 Dams Conflicts

After introduction of Dam building concept in India in a similar to western countries, the problem of ecological disruption and social conflicts started in India. These social conflicts and disruption are more aggravated in India than the havoc caused in the Western countries because India is a riparian civilization. Most of India's river valleys are highly populated and rivers have provided the primary life-support systems for our riparian settlements. Large dams, intensive irrigation and large diversions in India resulting in three different types of conflicts. The first type is related to large-scale displacement and uprooting of people from their ancestral homelands leading to ecological refugees. This conflict, which originally expressed itself through human rights struggles based on the violation of rights of displaced people, has now taken an ecological turn, with human rights issues being perceived as intimately linked with ecological issues.

The second type of conflict related to water projects arises from the ecological impact of impounding large quantities of water, transporting it across drainage boundaries and using it for intensive irrigation. Changes in water flows create changes upstream as well as downstream. Such changes generate conflicts not merely between the people and the state, but also between different communities and different states. The third type of conflict arising mainly from large river diversions over water rights. Interests of people of different regions are articulated through regional governments, and regional conflicts take the form of inter-state conflicts over the sharing of river waters.

Case study

The Krishna river, one of the most important rivers of South India, was chosen for the ecological analysis of conflicts over river waters since it traverses through the most arid and drought prone regions in South India and there are intense and diverse demands for its water from different regions for diverse uses.

1.5 Drought

Drought is recognized as one of the most insidious causes of human misery. It has today the unfortunate distinction of being the natural disaster that annually claims the most victims. Its ability to cause widespread misery is actually increasing.

While generally associated with semiarid climates, drought can occur in areas that normally enjoy adequate rainfall and moisture levels. In the broadest sense, any lack of water for the normal needs of agriculture, livestock, industry, or human population may be termed a drought. The cause may be lack of supply, contamination of supply, inadequate storage or conveyance facilities, or abnormal demand. Drought, as commonly understood, is a condition of climatic dryness that is severe enough to reduce soil moisture and water below the minimum necessary for sustaining plant, animal, and human life. Drought is usually accompanied by hot, dry winds and may be followed by damaging floods.

1.5.1 Regions of drought in the world

Droughts occur in all of the world's continents. In recent decades the most severe and devastating to human populations have been in Africa, perhaps giving the impression that droughts are principally an African problem. In fact devastating droughts have occurred in virtually all of the major semiarid regions of the world as well as in many zones that are normally temperate climates with significant annual rainfalls.

In addition to the droughts in the African Sahel, there have recently been major droughts in northeast Brazil, Chile, Ethiopia, the Philippines, the Bolivian altiplano, and India. Near-drought conditions were reported in the grain-producing regions of the United States and the Soviet Union affecting greatly the international food supply and demand. Trends in the occurrence of droughts indicate they are becoming more frequent on the edges of desert lands and where agricultural, lumbering, and livestock grazing practices are changing.

1.5.2 Effects of Drought

If a drought is allowed to continue without proper measure, the impact on development can be severe. Food shortages may become chronic.

If drought response is treated as only a relief operation, it may wipe out years of development work, especially in rural areas. Agricultural projects in particular are most likely to be affected by droughts. For those in agricultural development, droughts or the threat of droughts should be considered a part of the overall development equation.

1.6 Flood

Flood is generally defined as any unusual high stream flow that overtops the natural or artificial banks of a stream and it is a natural characteristic of rivers. The floodplains are normally dry land areas. They are an integral part of a river system that acts as a natural reservoir and temporary channel for flood waters. If more runoff is generated beyond the capacity of stream channel can accommodate, the water will overtop the stream banks and spread over the floodplain. The ultimate effect of flooding is damage the extent of which depends on how high the water goes above normal restraints or embankments. Furthermore, floods can form due to abnormal heavy precipitation falls on flat terrain at such a rate that the soil cannot absorb the water or the water cannot run off as fast as it falls.

1.6.1 Effects of Flood

Among all the disasters, flood disasters affect the most people next to economical the damage extent of droughts. But now days the number of area affected by floods is increasing much more rapidly than the number of regions suffering from droughts due to increasing measure taken to manage the drought situation. In fact, flooding is one natural hazard that is becoming a greater threat rather than a constant or declining one.

1.6.2 Causes of Flood

Besides raining, floods are also caused by human changes to the surface of the earth. Farming, deforestation, and urbanization increase the runoff from rains; thereby storms that previously would have caused no flooding today inundate vast areas. More recently reckless building in vulnerable areas, poor watershed management, and failure to control the flooding also help create the disaster condition. Ecologists have recently found evidence that human endeavors may directly be affecting the weather conditions that produce extensive and heavy rains. Irrigation of dry lands creates moisture conditions that contribute to increased humidity and evaporation, which in turn lead to increased rainfall. This situation is experienced especially in desert areas where large lakes are built to provide water either for irrigation or for nearby settlements.

1.7 Environmental Effects of Mineral Resources

(i) Environmental Effects of Solid Mineral and Oil Mining

Due to mainly open cast nature, the mining of Solid Minerals in India negatively impacts on the Environment. In the Tamil Nadu, Andhra , Bihar states area for instance, one can see the various pits due to open cast mining. The stagnant waters breed frogs and mosquitoes which could pose health hazards.

Mining impacts can be sub-divided for convenience into four categories: the effects of the mine itself, the disposal of mine wastes, the transport of the mineral, and the processing of the ore which often involves or produces dangerous materials. These activities may take place at the same site, in which case the impacts are combined, or they may be separated by considerable distances. The effects of each are discussed separately here.

Direct Impact of the Mine

Many mines are at the surface (sometimes called open cast mines). They may consist of a great hole in the ground (most copper mines), strip mining along the tops of ridges (nickel in New Caledonia), the digging out of ore in pits and sheets across a large surface (phosphates), or a pit or hillside excavation (rock and sand quarries). With such mines, the land surface over a considerable area is destroyed, and what is left behind may be unstable, producing landslides, erosion, siltation, and polluted water. Such land is generally useless after the mining ends, and may continue to cause environmental problems long after the mine has closed.

When the mine is beneath the surface and is composed of tunnels, the disturbance is usually localized at the mine entrance unless there are later cave-ins or some interference with ground-water supplies.

The drilling of oil wells may cause some local environmental effects from the machinery and drilling muds used, and there is a slight risk of blow-outs or accidental breakage leading to oil spillage, but once the well is in operation, there is generally little further effect.

Any kind of mine may cause some water pollution through water drained or pumped from the mine.

Disposal of Mine Wastes

In most mining operations, a great deal of waste rock and tailings must be moved to reach the ore. In addition to rock covering the ore or mixed in with the ore, there may be low grade ores that are not worth treating. The disposal of these wastes from the mine often causes more environmental damage than the mine itself. The problems come both from the enormous quantities of rock to be disposed of, and the fact that some mine wastes may be toxic to plants or cause water pollution as water drains through them.

The dumping of these wastes may affect more land surface than the mine. Waste dumps are subject to erosion, and are difficult to stabilize or revegetate. It is sometimes even necessary to build dams to hold the wastes back. Inadequately contained mine wastes have damaged or destroyed rivers and agricultural land and caused heavy flooding. In some mines the wastes are intentionally washed in to a river so that they will be carried away to the sea, with severe effects on the river, its valley, and the sea coast at the river mouth.

Transport of Ores

Mines are usually located in remote areas, requiring the transport of the mineral in some form from the mine to a port where it can be shipped to its markets. The infrastructure for the transport of large quantities of heavy material may involve roads, railroads, pipelines and conveyor belts, the construction and operation of which may have considerable environmental impacts, particularly since the terrain crossed is often difficult. Major port facilities and storage areas are also generally required.

Ore Treatment or Refining

Since transportation cost in island regions are very high, and are related to the bulk of the material, there is a strong incentive for at least some treatment of the ore to concentrate it if not refine it completely. This generally requires large industrial installations and the further production of wastes to be disposed of. Furthermore, ore treatment often requires dangerous chemicals like mercury or cyanide, or strong acids or alkalis, and the metals being treated may themselves be toxic. Some are even known to accumulate in food chains, threatening human food supplies. Even if the wastes are treated, there usually is some spillage, and there is always the risk of accidents.

Special Impacts in Coastal Areas

Where mines occur in coastal areas, or mine wastes are released there, additional environmental impacts can be expected on sensitive coastal resources. Corals and other forms of life on the bottom generally require clear water for growth, and are easily smothered by any input of tailings or other sediment. Dredging and other forms of coastal mining thus both destroy the immediate site and affect a much larger area through the sediment stirred up in the water. Removing sand and gravel from the beach or lagoon bottom can also cause beach disappearance and serious coastal erosion. Copper and other toxic materials from mines may kill most marine life in the vicinity of treatment and shipping facilities.

Long-term Problems

Mines as development projects can create employment and provide a significant source of national income and foreign exchange. However, as a development based on a non-renewable resource, mines are only viable for a limited period and are then abandoned, requiring the writing off of all the capital investment involved. Many islands have mining ghost towns and rusting equipment that are the only remaining signs of brief periods of prosperity. When the mine closes, there is generally no one left to be legally or financially responsible for the long-term effects. Those costs are left to the government and people. Often all that remains of the mine are the environmental problems and large areas of ruined land on islands that can ill afford to waste limited land resources.

The solution to such long-term problems is to build environmental controls and provision for rehabilitation into the project from the beginning.

Environmental Effects of Fossile Fuel

Coal, oil, and gas consist largely of carbon and hydrogen. The process that we call "burning" actually is chemical reactions with oxygen in the air. For the most part, the carbon combines with oxygen to form carbon dioxide (CO_2), and the hydrogen combines with oxygen to form water vapor (H_2O). In both of these chemical reactions a substantial amount of energy is released as heat. Since heat is what is needed to instigate these chemical reactions, we have a chain reaction: reactions cause heat, which causes reactions, which cause heat, and so on. Once started the process continues until nearly all of the fuel has gone through the process (i.e., burned), or until something is done to stop it. Of course, the reason for arranging all this is to derive the heat.

The carbon dioxide that is released is the cause of the greenhouse effect we will be discussing. A large coal-burning plant annually burns 3 million tons of coal to produce 11 million tons of carbon dioxide. The water vapor release presents no problems, since the amount in the atmosphere is determined by evaporation from the oceans if more is produced by burning, that much less will be evaporated from the seas.

1.8 Food Rseources

Food security depends on available world supplies of food, the income of the designated population, accessibility to the available supplies, the consumption rate of food, and the amount that can be set aside for future use. Global food supply has improved enormously since the early 1960s. World food and agricultural production has never experienced more favorable conditions than in the 1980s and 1990s. The agricultural sector on average has kept up with population growth and demand for agricultural produce. Agricultural production has increased and world food supplies are 18% higher than 30 years ago.

The world grain harvest in 1994, at 1,747 million tons, was up 2.9% from the de-pressed 1993 harvest of 1,697 million tons. World grain production (mainly wheat, corn, and rice) has shown an upward trend, with the exception of slight fluctuations in some years due primarily to drought and other natural disasters. In the decade of the 1990s, global grain yields per hectare were nearly 2.5 times the 1.15 tons per hectare of the 1930s

World meat production has shown an increasing trend with 184 million tons in 1994, up from 177 million tons in 1993. The impressive gain in overall meat output boosted per capita production to 33 kilograms, the highest ever achieved. Pork continued to widen its lead over beef as the world's most popular meat, and poultry production continued to be the most dynamic subsector of the meat industry. These increases in meat production are consequences of improvement in income levels, since meat becomes a more important component in diets as income increases.

Thus, it is evident that, more than ever before, an adequate amount of food is being produced today. Based on current trends and improved agricultural technology, world aggregate food production will be sufficient to meet demand in the decades ahead. By 2010, it is anticipated that world food supplies will be adequate to feed the global population. Right now there are viable alternative technologies in agriculture to secure the world food supply.

A report on analysis of food security and world trade prospects, predicts that the world food situation will continue to improve, at least over the next two to three decades; there will be continued improvements in productivity, the rate of growth of supply will exceed that of demand, and real international prices of food will also decline. Johnson further contends that the real per capita incomes of the majority of the populations in developing countries will continue to increase, con-tributing to an improvement in food security.

This optimistic forecast, however, does not mean that in every country food security will improve. It is evident that there is more than enough food for every man, woman, and child on the planet, but all too often the poor do not have access to that food. Des-pite the availability of viable technologies to increase food and agricultural production, economic and social progress is not expected to occur at similar rates across countries. This is because many of the poorer countries are unable to be self-sufficient in food and agricultural production due to various economic, social, and political constraints.

1.8.1 Technological Improvements in Food Production

World food shortages can be eliminated by increasing food and agricultural production through the application of modern technology. Also supplying modern inputs—such as large-scale irrigation, chemical fertilizers, farm machinery, and pesticides—can improve the productive capacity of the land. However, when a new agricultural technology enters a system characterized by unequal power relationships, it brings greater profits only to those who already have some combination of land, financial resources, creditworthiness, and political influence.

For example, a research study completed by the International Labor Organization (ILO) showed that in the South Asian countries (Pakistan, India, Thailand, Malaysia, the Philippines, and Indonesia), where the focus was on increased agricultural production and where the gross national product (GNP) has risen, the majority of the rural population was worse off than ever before.

In many cases, the structure of production agriculture has also created an hindrance for increased food production. An increasing dependence of the food supply on stocks of fossil energy sources that are being rapidly depleted is frequently ignored by economists and agronomists in their praise of the achievements of technological progress in food and agricultural production. Improved technology cannot substitute for shortage of essential natural resources. But, technology can help by promoting better use of the natural resources.

There has been insufficient research on appropriate technologies adapted to the agro-climatic conditions of many developing countries. New technology can have positive effects on food and agricultural production only if appropriate technology evolves within the framework of existing agricultural methods of production by first understanding how these traditional and social institutions and economic systems operate. Moreover, maximizing yield through modern intensive technologies usually requires imported capital, a resource very scarce in poverty-stricken countries. Even if agricultural production increases as a result of new technology, most of the agri-cultural production will go to pay for the imported capital. Thus, while use of imported technologies is an important factor in increasing food production and reducing hunger, it is not sufficient and has costs as well as benefits.

1.8.2 Environmental Issues Related to Increased Production

Poverty and environmental degradation are closely linked, often in a self-perpetuating negative spiral in which poverty accelerates environmental degradation, and degradation results in or exacerbates poverty while simultaneously reducing the productivity of the resource, and hence food production. While poverty is not the only cause of environmental degradation, it does pose the most serious environmental threat in many low-income countries. This problem is further heightened because many millions of people, living near the subsistence minimum, have exploited natural resources with inappropriate technologies in order to survive. The very success of export-oriented agriculture undermines both the position of the poor as well as the environment. When export commodity prices rise, small self-sufficient farmers are pushed off onto marginal land by cash crop producers seeking to profit from the higher commodity prices.

Generally, it is not only the quest for increased food production that threatens to destroy the environment; the damage to the environment is often inflicted by commercial cropping patterns of the large farms that export nonfood crops while pushing the rural majority to eke out a meager living on marginal lands. Thus, unless collaborative action is implemented to rapidly increase productivity, many more subsistence farmers living on the edge of poverty will be moving onto marginal lands, causing a considerable amount of natural resource degradation and ultimately leading to environmental damage.

1.9 Overgrazing and Its Effects

Domestic animals are reared for many purposes. Milk is consumed directly or used to make milk products like butter and yogurt, which are sold in the markets. The meat is another item of consumption. The hide is also used, and in the case of the sheep, the wool is a commodity of great commercial value. Even the animal dung is useful as manure. And most importantly, the animals are beasts of burden and also used on the farms.

Overgrazing by domestic animals has adverse effects on the vegetation – in the case of both alpine grasslands and forests. Due to this, the forests and the grasslands become bare and subsequently prone to soil erosion. The animals graze during spring, when the seedlings of various tree species, grasses and herbs are growing. This leads to the problems of regeneration as the future crop is adversely affected.

Seedlings of various species can get crushed and trampled under the hooves of cattle. In some cases the roots of a tree species may be exposed by trampling. This results in the death of the desired species. Selective grazing tends to alter the composition of the forest or grassland ecosystem, causing an increase in the population of undesired species, which are not consumed by the animals.

Indiscriminate grazing leads to the degradation of the soil. The soil becomes compact and the porosity is reduced. Soil aeration is also adversely affected. This in turn results in the seeds not germinating as they do not get sufficient air and water from the soil. Moreover, the hooves of the animals break down the soil aggregates, which gives a crumbly structure to the soil. Due to this, the soil loses its ability to absorb sufficient amounts of water. This translates into little percolation of rain water, most of which gets drained away. Since indiscriminate grazing leaves large parts of the land bare, the rainwater that gets drained away tends to erode the soil in its path. This degrades the soil further.

1.10 Impacts of Intensive Farming on Soil and Water Resources

Two major impacts are realized due to intensive farming, they are soil erosion and water contamination.

Soil Erosion

Soil erosion from farmland threatens the productivity of agricultural fields and causes a number of problems elsewhere in the environment. An average of 10 times as much soil erodes from American agricultural fields as is replaced by natural soil formation processes. Because it takes up to 300 years for 1 inch of agricultural topsoil to form, soil that is lost is essentially irreplaceable. The consequences for long-term crop yields have not been adequately quantified. The amount of erosion varies considerably from one field to another, depending on soil type, slope of the field, drainage patterns, and crop management practices;

and the effects of the erosion vary also. Areas with deep organic loams are better able to sustain erosion without loss of productivity than are areas where topsoils are shallower.

Erosion affects productivity because it removes the surface soils, containing most of the organic matter, plant nutrients, and fine soil particles, which help to retain water and nutrients in the root zone where they are available to plants. The subsoils that remain tend to be less fertile, less absorbent, and less able to retain pesticides, fertilizers, and other plant nutrients. Even when soil erosion is not excessive, intensive agriculture can impair soil quality by depleting the natural supplies of trace elements and organic matter. In natural ecosystems, soil fertility is maintained by the diverse contributions and recycling of nutrients by a wide range of plant and animal species. When this diversity is replaced by a single species grown year after year, some trace elements are depleted if not replaced by fertilization. The organic content of the soil also diminishes unless crop residues or other organic materials are supplied in sufficient quantities to replace that consumed over time.

Water Contamination

Contamination of Water: In the Northeast water supplies are generally plentiful, but are increasingly becoming threatened by contamination. Farming is one potential source of such contamination. Surface runoff carries manure, fertilizers, and pesticides into streams, lakes, and reservoirs, in some cases causing unacceptable levels of bacteria, nutrients, or synthetic organic compounds. Similarly, water percolating downward through farm fields carries with it dissolved chemicals, which can include nitrate fertilizers and soluble pesticides. In sufficient quantities these can contaminate groundwater supplies.

Fertilizers : Nutrients are lost from agricultural fields through runoff, drainage, or attachment to eroded soil particles. The amounts lost depend on the soil type and organic matter content, the climate, slope of the land, and depth to groundwater, as well as on the amount and type of fertilizer and irrigation used.

1.11 Effects of Agricultural Chemicals on The Environment

Two chemicals namely Fertilizer and Pesticides are causing major threats to the environment in the form of water pollution and their effects are as follows

(i) Fertilizer

The three major nutrients in fertilizers are nitrogen, phosphorus, and potassium. Of these, nitrogen is the most readily lost because of its high solubility in the nitrate form. Leaching of nitrate from agricultural fields can elevate concentrations in underlying groundwater to levels unacceptable for drinking water quality.

Phosphorus does not leach as readily as nitrate because it is more tightly bound to soil particles. However, it is carried with eroded soils into surface water bodies, where it may cause excessive growth of aquatic plants. If this process proceeds far enough, lakes and reservoirs become choked with decaying mats of algae, which have offensive odors and can cause fish kills from the resulting lack of dissolved oxygen.

Potassium, the third major nutrient in fertilizers, does not cause water quality problems because it is not hazardous in drinking water and is not a limiting nutrient for growth of aquatic plants. It is tightly held by soil particles and so can be removed from fields by erosion, but generally not by leaching.

(ii) Pesticides

Pesticides, the trend toward intensive crop production in modern farming has led to increased potential for damage by pests and diseases. Predators that would be present in a mixed biological community are not supported by large fields of a single crop; so farmers, instead, rely on chemical measures for crop protection. Usage of pesticides in farms all over the world has risen ten fold over the past 40 years as agriculture has become more intensive. One drawback to this is that pesticides generally kill not only the pest of concern, but also a wide range of other organisms, including beneficial insects and other pest predators. Once the effect of the pesticide wears off, the pest species is likely to recover more rapidly than its predators because of differences in the available food supply. Previously unimportant species may also become significant crop pests when their natural predators are killed by pesticide applications.

Pesticides and pesticide residues can be transported by erosion and runoff to off-site areas in ways similar to those operative in the case of nutrients. These losses are determined by the persistence of the material, the strength with which it is adsorbed to sediments, and its solubility in water. Although data on a national scale are quite limited, reported pesticide concentrations in streams and lakes are generally very low. The delivery ratio for pesticides to surface waters is generally less than 5 percent.

Many people, for various reasons, are apprehensive about situations involving synthetic chemicals as contaminants in the water supply or food chain. It is becoming increasingly evident that these chemicals will be very difficult to keep out of the water or food chain if they are used extensively in crop production. Intensive monitoring and predictive programs are helpful but not foolproof, and these programs are very expansive.

1.12 Energy Sources

There are two energy sources namely Non-renuwable and renewable. These are as follows.

Renewable Energy

Renewable energy sources capture their energy from existing flows of energy, from on-going natural processes, such as sunshine, wind, flowing water, biological processes, and geothermal heat flows. Generally renewable energy is from an energy resource that is replaced rapidly by a natural process such as power generated from the sun or from the wind.

Various forms of Renewable Energy

Wind Energy

Winds are formed due to uneven heating of earth due to sun. The kinetic energy in the wind can be used to run wind turbine with capacity of producing 5 MW of power. The power output is a function of the cube of the wind speed, so such turbines generally require a wind in the range 5.5 m/s (20 km/h), and in practice relatively few land areas have significant prevailing winds. There are now many thousands of wind turbines operating in various parts of the world, with utility companies having a total capacity of 59,322 MW. Capacity in this case means maximum possible output which does not count load factor.

New wind farms and offshore wind parks are being planned and built all over the world. This has been the most rapidly-growing means of electricity generation at the turn of the 21st century and provides a complement to large-scale base-load power stations. Most deployed turbines produce electricity about 25% of the time (load factor 25%), but some reach 35%. The load factor is generally higher in winter. It means that a 5 MW turbine can have average output of 1.7 MW in the best case.

Global winds long-term technical potential is believed to be 5 times current global energy consumption or 40 times current electricity demand. This requires 12.7% of all land area, or that land area with Class 3 or greater potential at a height of 80 meters. It assumes that the land is covered with 6 large wind turbines per square kilometer. Offshore resources experience mean wind speeds of ~90% greater than that of land, so offshore resources could contribute substantially more energy. This number could also increase with higher altitude ground based or airborne wind turbines.

Wind strengths vary and thus cannot guarantee continuous power. Some estimates suggest that 1,000MW of wind generation capacity can be relied on for just 333 MW of continuous power. While this might change as technology evolves, advocates have suggested incorporating wind power with other power sources, or the use of energy storage techniques, with this in mind. It is best used in the context of a system that has significant reserve capacity such as hydro, or reserve load, such as a desalination plant, to mitigate the economic effects of resource variability.

Wind power is renewable and is one of the few energy sources that contributes to green house gas mitigation, because it removes energy directly from the atmosphere without producing net emissions of greenhouse gases such as carbon dioxide and methane (others greenhouse gas mitigating energy sources include solat thermal and ocean thermal).

Water Power

Energy in water can be harnessed and used, in the form of motive energy or temperature differences. There are many forms as follows

(i) Hydroelectric, energy a term usually reserved for hydroelectric dams.

(ii) Tidal power which captures energy from the tides in horizontal direction. Tides come in, raise waterlevels in a basin, and tides roll out. The water must pass through a turbine to get out of the basin.

(iii) Tidal stream power , which does the same vertically, capturing the stream of water as it is pushed around the world by the tides.

(iv) Wave power,which uses the energy in waves. The waves will usually make large pontoons go up and down in the water.

(v) Ocean thermal energy conversion, which uses the temperature difference between the warmer surface of the ocean and the cool (or cold) lower recesses. To this end, it employs a cyclic heat engine.

(vi) Deep lake water cooling, not technically an energy generation method, though it can save a lot of energy in summer. It uses submerged pipes as a heat sink for climate control system. Lake-bottom water is a year-round local constant of about 4 ° C.

(vii) Blue energy,the reverse of desalination. A difference in salt concentration exists between seawater and river water. This gradient can be utilized to generate electricity by separating positive and negative ions by ion specific membranes. Brackish water is produced. This form of energy is still very expensive.

Solar Energy

Solar power can be used to:

(i) generate electricity using solar cells

(ii) generate electricity using concentrated solar power

(iii) heat buildings, directly
(iv) heat buildings, through heat pumps
(v) heat foodstuffs, through solar ovens.

Usually the sun does not provide constant energy to any spot on the Earth, so its use is limited. Solar cells are often used to power batteries, as most other applications would require a secondary energy source, to cope with outages. Some homeowners use a solar system which sells energy to the grid during the day, and draw energy from the grid at night; this is to everyone's advantage, since power demand for air conditioning is highest during the day. Concentrated solar power plants work best in hot deserts and other places with plenty of direct sunshine.

Geothermal Energy

Geothermal energy ultimately comes from radiactive decay in the core of the Earth, which heats the Earth from the inside out, and from the sun, which heats the surface. It can be used in three forms

(i) Geothermal electricity
(ii) Geothermal heating, through deep Earth pipes
(iii) Geothermal heating, through a heat pump.

Usually, the term 'geothermal' is reserved for thermal energy from within the Earth.

Geothermal electricity is created by pumping a fluid (oil or water) into the Earth, allowing it to evaporate and using the hot gases vented from the earth's crust to run turbine linked to electrical generators.

The geothermal energy from the core of the Earth is closer to the surface in some areas than in others. Where hot underground steam or water can be tapped and brought to the surface it may be used to generate electricity. Such geothermal power sources exist in certain geologically unstable parts of the world such as Iceland, New Zealand, United states, The Philippines, and italy. Iceland produced 170 MW geothermal power and heated 86% of all houses in the year 2000 through geothermal energy. Some 8000 MW of capacity is operational in total.

Geothermal heat from the surface of the Earth can be used on most of the globe directly to heat and cool buildings. The temperature of the crust a few feet below the surface is buffered to a constant 7 to 14 °C (45 to 58 °F), so a liquid can be pre-heated or pre-cooled in underground pipelines, providing free cooling in the summer and, via a heat pump, heating in the winter. Other direct uses are in agriculture (greenhouses), aquaculture and industry.

Although geothermal sites are capable of providing heat for many decades, eventually specific locations cool down. Some interpret this as meaning a specific geothermal location can undergo depletion. Others see such an interpretation as an inaccurate usage of the word depletion because the overall supply of geothermal energy on Earth, and its source, remain nearly constant. Geothermal energy depends on local geological instability, which, by definition, is unpredictable, and might stabilise.

Biomass

Plants use photosynthesis to store solar energy in the form of chemical energy . Biofuel is any fuel that derives from biomass, including living organisms or their metabolic byproducts, such as cow manure.

Typically biofuel is burned to release its stored chemical energy. Research into more efficient methods of converting biofuels and other fuels into electricity utilizing fuel cells is an area of very active work.Biomass, also known as biomatter, can be used directly as fuel or to produce liquid biofuel. Agriculturally produced biomass fuels, such as bidiesel, ethanol and bagasse (often a by-product of sugar cane cultivation) can be burned in internal combustion engine or boilers.

A drawback is that all biomass needs to go through some of these steps: it needs to be grown, collected, dried, fermented and burned. Biomatter energy, under the right conditions, is considered to be renewable.

Liquid Biofuel

Liquid biofuel is usually bioalcohol such as ethanol and biodiesel. Biodiesel can be used in modern diesel vehicles with little or no modification and can be obtained from waste and crude vegetable and animal oil and fats (lipids). A major benefit of biodiesel is lower emissions. The use of biodiesel reduces emission of carbon monoxide and other hydrocarbons by 20 to 40 percent. In some areas corn, sugarbeets, cane and grasses are grown specifically to produce ethanol (also known as alcohol) a liquid which can be used in internal combustion engine and fuell cells. Ethanol is being phased into the current energy infrastructure. E85 is a fuel composed of 85% ethanol and 15% gasoline that is currently being sold to consumers.

In the future, there might be bio-synthetic liquid fuel available. It can be produced by Fishcer-Tropsch processes, also called Biomass-To-Liquids (BTL).

Solid Biomass

Direct use is usually in the form of combustible solids, either firewood or combustible field crops. Field crops may be grown specifically for combustion or may be used for other purposes, and the processed plant waste then used for combustion. Most sorts of biomatter, including dried manure, can actually be burnt to heat water and to drive turbines.

Sugar cane residue, wheat chaff, corn cobs and other plant matter can be, and is, burnt quite successfully. The process releases no net CO_2.

Solid biomass can also begasifed, and used as described in the next section.

Biogas

Many organic materials can release gases, due to metabolisation of organic matter by bacteria (fermentation). Landfills actually need to release this gas to prevent dangerous explosions. Animal feces releases methane under the influence of anaerobic bacteria. Also, under high pressure, high temperature, anaerobic conditions many organic materials such as wood can be gasified to produce gas. This is often found to be more efficient than direct burning. The gas can then be used to generate electricity and/or heat.

Biogas can easily be produced from current waste streams, such as: paper production, sugar production, sewage, animal waste and so forth. These various waste streams have to be slurried together and allowed to naturally ferment, producing methane gas. We just need to convert current sewage plants to biogas plants, build more locally centered smaller biogas plants and plan for the future. Biogas production has the capacity to provide us with about half of our energy needs, either burned for electrical productions or piped into current gas lines for use. It just has to be done and made a priority. Besides, when a plant has extracted all the methane it can, we are left with a better fertilizer for our farms than we started with.

Renewable natural gas is a biogas which has been upgraded to a quality similar to natural gas. By upgrading the quality to that of natural gas, it becomes possible to distribute the gas to the mass market via the existing gas grid.

Small Scale Energy Sources

There are many small scale energy sources that generally cannot be scaled up to industrial size. A short list:

(i) Piesolectric crystals generate a small voltage whenever they are mechanically deformed. Vibration from engines can stimulate piezoelectric crystals, as can the heels of shoes

(ii) Some watches are already powered by kinetics, in this case movement of the arm

(iii) Electrokinetics generate electricity from the kinetic energy in water that is pumped through tiny channels

(iv) Special antennae can collect energy from stray radio waves or theoretically even light (EM radiation).

Non Renewable Energy Resources

Fossil fuels

Coal, crude and crude based products are falling in this type. This wil be depleting in nature. In addition this will be have a lot enviromental issues like green house effect, global warming, and acid rain etc. Earth's sources would be depleted within some 50 years. Since then, large deposits of deep-Earth oil have been found, which has extended this timetable. Because the current rate of consumption exceeds the rate of renewal (if, indeed, there is renewal of fossil fuels), the Earth will eventually run out of fossil fuels.

1.13 Waterlogging

Water logging is the problem associated with excessive irrigation on poorly drained soils. This occurs in poorly drained soils where water can't penetrate deeply. For example, there may be an impermeable clay layer below the soil. It also occurs on areas that are poorly drained topographically. The irrigation water (and/or seepage from canals) eventually raises the water table in the ground – the upper level of the groundwater – from beneath. Growers don't generally realize that waterlogging is happening until it is too late – tests for water in soil are apparently very expensive.

The raised water table results in the soils becoming waterlogged. When soils are water logged, air spaces in the soil are filled with water, and plant roots essentially suffocate – lack oxygen. Waterlogging also damages soil structure. Worldwide, about 10-15 % of all irrigated land suffers from water logging. This is an area about the size of Idaho. As a result, productivity has fallen about 20-25 % in this area of cropland. Both water logging and salinization could be reduced if the efficiency of irrigation systems could be improved, and more appropriate crops (less water hungry) could be grown in arid and semi-arid regions.

1.14 Land Resources

"Land is a delineable area of the earth's terrestrial surface, encompassing all attributes of the biosphere immediately above or below this surface including those of the near-surface climate the soil and terrain forms, the surface hydrology (including shallow lakes, rivers, marshes, and swamps), the near-surface sedimentary layers and associated groundwater reserve, the plant and animal populations, the human settlement pattern and physical results of past and present human activity (terracing, water storage or drainage structures, roads, buildings, etc.)."

(i) This definition conforms to land system units. landscape-ecological units or unites de terroir, as building blocks of a watershed (catchment area) or a phytogeographic unit (biome).

(ii) The definition of a natural land unit as defined above is distinctive from an administrative unit of land (territoire) which can be of any size (individual holding, municipality, province, state, etc.) and which normally encompasses a number of natural units or parts of them.

(iii) The components of the natural land unit can be termed land resources, including physical, bionic, environmental, infrastructural, social and economic components, inasmuch as they are fixed to the land unit.

(iv) Included in the land resources are surface and near-surface freshwater resources. Part of these move through successive land units, but then the local flow characteristics can be considered as part of the land unit. The linkages between water and land are so intimate at the management level that the water element cannot be excluded (land as a unit intermixed with water, with its land use in part depending on access to that water, and the unit at the same time affecting the quality and quantity of the passing water). Only the freshwater harnessed in major reservoirs outside the natural land unit, or pumped from rivers at upstream sites, can be considered as a separate resource.

(v) Underground geological resources (oil, gas, ores, precious metals), and deeper geohydrological resources that normally bear no relation to the surface topography such as confined aquifers, are excluded from the group of components of the natural land unit, although it is recognized that some countries consider them as part of individual land ownership (and hence with rights to exploit or sell them).

In short the natural unit of land has atmospheric climate down to groundwater resources, repetitive sequence of soil, terrain, hydrological, and vegetative or land use elements

1.15 Land Degradation

Land degradation is a human induced or natural process which negatively affects the capacity of land to function effectively within an ecosystem. Desertification is land degradation occurring in arid, semiarid and dry subhumid areas of the world.

Land degradation causes losses to agricultural productivity in many parts of the world.

The causes of land degradation are mainly anthropogenic and mainly agriculture related. The major causes are :

(i) Land clearing and deforestation

(ii) Agricultural depletion soil nutrients

(iii) Urban conversion

(iv) Irrigation

(v) Pollution

(vi) Erosion by wind and water

(vii) Soil acidification or alkalinization

(viii) Salination

(ix) Destruction of soil structure including loss of organic matter

Severe land degradation affects a significant portion of the earth's arable lands, decreasing the wealth and economic development of nations.

General public often assume that land degradation only affects soil productivity. But, the effects of land degradation often have more significant impacts on receiving water courses (rivers, wetlands and lakes) since soil, along with nutrients and contaminants associated with soil, are delivered in large quantities to environments that respond detrimentally to their input. Land degradation therefore has potentially disasterous impacts on lakes and reservoirs that are designed to alleviate flooding, provide irrigation, and generate Hydro-Power.

1.16 Soil Erosion

Erosion is an intrinsic natural process, but in many places it is increased by human land use. Poor land use practices include deforestation, overgrazing, unmanaged construction activity and road or trail building. However, improved land use practices can limit erosion using techniques like terrace-building and tree planting.

1.16.1 Types of Erosion

There are five types of erosion (i) Gravity (ii) Water (iii) Shoreline (iv) Ice and (v) Wind erosion. These are explained as follows:

(i) Gravity Erosion

Mass wasting also known as mass movment is the down-slope movement of rock and sediments, mainly due to the force of gravity. Mass wasting is an important part of the erosional process, as it moves material from higher elevations to lower elevations where transporting agents like streams and glaciers can then pick up the material and move it to even lower elevations. Mass-wasting processes are occurring continuously on all slopes; some mass-wasting processes act very slowly, others occur very suddenly, often with disastrous results. Any perceptible down-slope movement of rock or sediment is often referred to in general terms as a landslide. However, landslides can be classified in a much more detailed way that reflects the mechanisms responsible for the movement and the velocity at which the movement occurs.

Movement of soil in single unit called slump happens on steep hillsides, occurring along distinct fracture zones, often within materials like clay, that, once released, may move quite rapidly downhill. They often will show a spoon-shaped depression within which the material has begun to slide downhill. In some cases the slump is caused by water beneath the slope weakening it. In many cases it is simply the result of poor engineering along highways where it is a regular occurrence.

Surface Creep is the slow movement of soil and rock debris by gravity which is usually not perceptible except through extended observation. However, the term can also describe the rolling of dislodged soil particles 0.5 to 1.0 mm in diameter by wind along the soil surface.

(ii) Water Erosion

Splash erosion is the detachment and airborne movement of small soil particles caused by the impact of raindrops on soil. Sheet erosion is the result of heavy rain on bare soil where water flows as a sheet down any gradient carrying soil particles. Where precipitation rates exceed soil infiltration rates, runoff occurs. Surface runoff turbulence can often cause more erosion than the initial raindrop impact. Gully erosion results where water flows along a linear depression eroding a trench or gully.

Valley or stream erosion occurs with continued water flow along a linear feature. The erosion is both downward, deepening the valley, and headward, extending the valley into the hillside. In the earliest stage of stream erosion the erosive activity is dominantly vertical, the valleys have a typical **V** cross-section, and the stream gradient is relatively steep. When some base level is reached the erosive activity switches to lateral erosion which widens the valley floor and creates a narrow floodplain. The stream gradient becomes nearly flat and lateral deposition of sediments becomes important as the stream meanders across the valley floor.

In all stages of stream erosion by far the most erosion occurs during times of flood when more and faster moving water is available to carry a larger sediment load.

(iii) Shoreline Erosion

Shoreline erosion, on both exposed and sheltered coasts, primarily occurs through the action of currents and waves, but sea level change can also play a role. Sediment is transported along the coast in the direction of the prevailing current (longshore drift). When the upcurrent amount of sediment is less than the amount being carried away, erosion occurs. When the upcurrent amount of sediment is greater, sand or gravel banks will tend to form. These banks may slowly migrate along the coast in the direction of the longshore drift, alternately protecting and exposing parts of the coastline.

(iv) Ice Erosion

Ice erosion is caused by movement of ice, typically as glaciers. Glaciers can scrape down a slope and break up rock and then transport it, leaving moraines, drumlins, and glacial erratics in their wake typically at the terminus or during glacier retreat. *Ice wedging* is the weathering process where water trapped in tiny rock cracks freezes and expands, causing the breakup of the rock. This can lead to gravity erosion on steep slopes. The scree which forms at the bottom of a steep mountainside is mostly formed from pieces of rock broken away by this means. It is a common engineering problem wherever rock cliffs are alongside roads and morning thaws can drop hazardous rock pieces onto the road.

(v) Wind Erosion

Wind erosion, also known as eolian erosion is the movement of rock and/or sediment by the wind. Windbreaks are often planted by farmers to reduce wind erosion. This includes the planting of trees, shrubs, or other vegetation, usually perpendicular or nearly so to the principal wind direction. The wind causes dust particles to be lifted and therefore moved to another region. Wind erosion generally occurs in areas with little or no vegetation, often areas where there is not enough rainfall to support vegetation.

1.17 Desertification

Desertification is the process which turns productive into non- productive desert as a result of poor land-management. Desertification occurs mainly in semi-arid areas (average annual rainfall less than 600 mm) bordering on deserts. Overgrazing is the major cause of desertification worldwide.

1.17.1 Causes of Desertification

(i) Cultivation of marginal lands, i.e lands on which there is a high risk of crop failure and a very low economic return, for example, some parts of South Africa where maize is grown.

(ii) Destruction of vegetation in arid regions, often for fuel wood.

(iii) Poor grazing management after accidental burning of semi-arid vegetation.

(iv) Incorrect irrigation practices in arid areas can cause salinization, i e the build up of salts in the soil, which can prevent plant growth.

If the above described activities coincide with drought, the rate of desertification increases dramatically. Increasing human population and poverty contribute to desertification as poor people may be forced to overuse their environment in the short term, without the ability to plan for the long term effects of their actions. Where livestock has a social importance beyond food, people sometimes will be reluctant to reduce their stock numbers.

1.17.2 Effects of Desertification

Desertification reduces the ability of land to support life, affecting wild species, domestic animals, agricultural crops and people. Desertification may lead to accelerated soil erosion by wind and water. South Africa is losing approximately 300-400 million tonnes of topsoil every year due to desertification. As vegetation cover and soil layer are reduced, rain drop impact and run-off increases.

Water is lost off the land instead of soaking into the soil to provide moisture for plants. Even long-lived plants that would normally survive droughts die. A reduction in plant cover also results in a reduction in the quantity of humus and plant nutrients in the soil, and plant production drops further. As protective plant cover disappears, floods become more frequent and more severe. Desertification is self-reinforcing, i.e. once the process has started, and conditions are set for continual deterioration.

1.18 Conservation of Natural Resources

Conservation, sustainable use and protection of natural resources including plants, animals, minerals, soils, clean water, clean air, and fossil fuels such as coal, petroleum, and natural gas is one of the important aspects in the present day situations. Natural resources are grouped into two categories namely renewable and nonrenewable. A renewable resource is one that may be replaced over time by natural processes, such as fish populations or natural vegetation, or is inexhaustible, such as solar energy. The goal of renewable resource conservation is to ensure that such resources are not consumed faster than they are replaced. Non-renewable resources are those in limited supply that cannot be replaced or can be replaced only over extremely long periods of time. Nonrenewable resources include fossil fuels and mineral deposits, such as iron ore and gold ore. Conservation activities for nonrenewable resources focus on maintaining an adequate supply of these resources well into the future.

Natural resources have to be conserved for their biological, economic, and recreational values, as well as their natural beauty and importance to local cultures. For example, tropical rain forests are protected for their important role in both global ecology and the economic livelihood of the local culture.

1.18.1 Conflicts

Conservation conflicts may arise when natural-resource shortages develop due to steadily increasing demands from a growing human population. Controversy frequently surrounds how a resource should be used, or allocated, and for whom. For example, a river may supply water for agricultural irrigation, habitat for fish, and water-generated electricity for a factory. Farmers, fishers, and industry people will try for unrestricted access to this river, but such freedom could destroy the resource, and therefore conservation methods are necessary to protect the river for future use.

Sometimes conflicts worsen when a natural resource crosses political boundaries. For example, the headwaters, or source, of a major river may be located in a different country than the country through which the river flows. There is no guarantee that the river source will be protected to accommodate resource needs downstream. In addition, the way in which one natural resource is managed has a direct effect upon other natural resources. Cutting down a forest near a river, may increases erosion, the wearing away of topsoil, and can lead to flooding. Eroded soil and silt cloud the river and adversely affect many organisms such as fish and important aquatic plants that require clean, clear freshwater for survival.

1.18.2 Challenges and Methods

The challenge of conservation is to understand the complex connections among natural resources and balance resource use with protection to ensure an adequate supply for future generations. To achieve this goal, a variety of conservation methods are generally used. Very widely used conservation methods are reducing consumption of resources; protecting them from contamination or pollution; reusing or recycling resources when possible; and fully protecting, or preserving, resources.

(i) Reduction of Consumption

Consumption of natural resources rises dramatically every year as the human population increases and standards of living rise. From 1950 to 2000 the world population more than doubled to 6 billion people, with nearly 80 percent living in developing, or poorer, nations. Besides the developed nations are equally responsible for the greatest consumption of natural resources because of their high standards of living. For instance, the average American consumes as much energy as 27 Filipinos or 370 Ethiopians. Conservation education and the thoughtful use of resources is necessary in the developed countries to reduce natural-resource consumption.

(ii) Protection from Contamination

To protect natural resources from pollution, individuals, industries, and governments have many obligations. These include prohibiting or limiting the use of pesticides and other toxic chemicals, limiting wastewater and airborne pollutants, preventing the production of radioactive materials, and regulating drilling and transportation of petroleum products. Failure to do so results in contaminated air, soil, rivers, plants, and animals. For example, if governments required that all oil tankers be fitted with double-layered hulls, the damages to fisheries and wildlife from the many oil spills of the 20th century, such as the 1967 Torrey Canyon oil spill in the English Channel, could have been reduced drastically.

(iii) Recycling

In many cases it is possible to reuse or recycle resources to reduce waste and resource consumption and conserve the energy needed to produce consumer products. For example, paper, glass, Freon, aluminum, metal scrap, and motor oil can all be recycled. A preventative measure called precycling, a general term for designing more durable, recyclable products such as reusable packaging, encourages reuse. It is advisable to establish a regulation for compulsory recycling laws in an attempt to reduce waste and consumption.

(iv) Preservation

Some resources are so unique or valuable that they are protected from activities that would destroy or degrade them. For example many parks and wilderness areas are protected from logging or mining in the United States because such activities would reduce the economic, recreational, and aesthetic values of the resource. Forests and wetlands can be protected from development because they enhance air and water quality and provide habitat for a wide variety of plants and animals. Unfortunately, these areas are often threatened with development because it is difficult to measure the economic benefits of cleaner air, cleaner water, and the many other environmental benefits of these ecosystems (the plants and animals of a natural community and their physical environment).

Test Your Understanding

1. Explain the forest resources in the world
2. Discuss the uses of Forest
3. Explain the Effects of deforestation
4. Mention the water resources and factors affecting them
5. Discuss briefly water conflicts with case study
6. Write short notes on the following
   (i) Drought
   (ii) Flood
   (iii) Environmental effects of mineral resources
   (iv) Technological improvement in food production
   (v) Overgrazing and its effects
7. What is water logging? Explain
8. Discuss briefly the causes of desertification
9. Discuss briefly about Conservation of Natural resources.

CHAPTER 2

Ecosystems and Biodiversity

2.1 Ecosystem

An ecosystem can be defined as a dynamic entity composed of a biological community and its associated abiotic environment. Often the dynamic interactions that occur within an ecosystem are numerous and complex. Ecosystems are also always undergoing changes or alterations to their biotic and abiotic components. Some of these alterations begin first with a change in the state of one component of the ecosystem which then cascades and sometimes amplifies into other components because of relationships.

The human activities of various kinds have resulted in caused a number of dramatic changes to a variety of ecosystems found on the Earth. Humans use and modify natural ecosystems through agriculture, forestry, recreation, urbanization, and industry. The most obvious impact of humans on ecosystems is the loss of biodiversity. The number of extinctions caused by human domination of ecosystems has been steadily increasing since the start of the Industrial revolution. The frequency of species extinctions is correlated to the size of human population on the Earth which is directly related to resource consumption, land-use change, and environmental degradation. Other human impacts to ecosystems include species invasions to new habitats, changes to the abundance and dominance of species in communities, modification of biogeochemical, and modification of hydrologic cycling, pollution, and climatic change.

2.1.1 Major Components of Ecosystems

Ecosystems are composed of a variety of abiotic and biotic components that function in an interrelated fashion. Some of the more important components are: *soil*, *atmosphere*, *radiation from the sun*, *water*, and *living organisms*.

Soils are much more complex than simple sediments. They contain a mixture of weathered rock fragments, highly altered soil mineral particles, organic matter, and living organisms. Soils provide nutrients, water, a home, and a structural growing medium for organisms. The vegetation found growing on top of a soil is closely linked to this component of an ecosystem through nutrient cycling.

The *atmosphere* provides organisms found within ecosystems with carbon dioxide for photosynthesis and oxygen for respiration. The processes of evaporation, transpiration, and precipitation cycle water between the atmosphere and the Earth's surface.

Solar radiation is used in ecosystems to heat the atmosphere and to evaporate and transpire water into the atmosphere. Sunlight is also necessary for photosynthesis. Photosynthesis provides the energy for plant growth and metabolism, and the organic food for other forms of life.

Most living tissue is composed of a very high percentage of *water*, up to and even exceeding 85 to 90 %. The protoplasm of a very few cells can survive if their water content drops below 10 %, and most are killed if it is less than 30-50 %. Water is the medium by which mineral nutrients enter and are translocated in plants. It is also necessary for the maintenance of leaf turgidity and is required for photosynthetic chemical reactions. Plants and animals receive their water from the Earth's surface and soil. The original source of this water is precipitation from the atmosphere.

Ecosystems made of a variety of living organisms with the classification of producers, consumer, or decomposer. Producers also called autotrophs are organism that can manufacture the organic compounds they use as sources of energy and nutrients. Most producers are green plants that can manufacture their food through the process of photosynthesis. Consumer also called heterotrophs get their energy and nutrients by feeding directly or indirectly on producers.

Usually there are two main types of consumers namely herbivorous and Carnivorous. Herbivores are consumers that can eat plants for their energy and nutrients whereas the organisms feeding on herbivores are called carnivores. Carnivores can also consume other carnivores. Plants and animals supply organic matter to the soil system through shed tissues and death. Consumer organisms that feed on this organic matter, or detritus, are known as detrivores or decomposer. The organic matter consumed by the detritivores is eventually converted back into inorganic nutrients in the soil and taken by plants as nutrients for the production of organic compounds. The Ecosystem model is shown in Fig.2.1 demonstrates the major components of the ecosystem and their interrelationships.

2.1.2 Energy and Matter Flow Concepts in Ecosystems

Many of the important relationships between living organisms and the environment are controlled by and large by the amount of available incoming energy received at the Earth's surface from the sun. The sun's energy which helps to drive biotic systems allows plants to convert inorganic chemicals into organic compounds. Only a very small

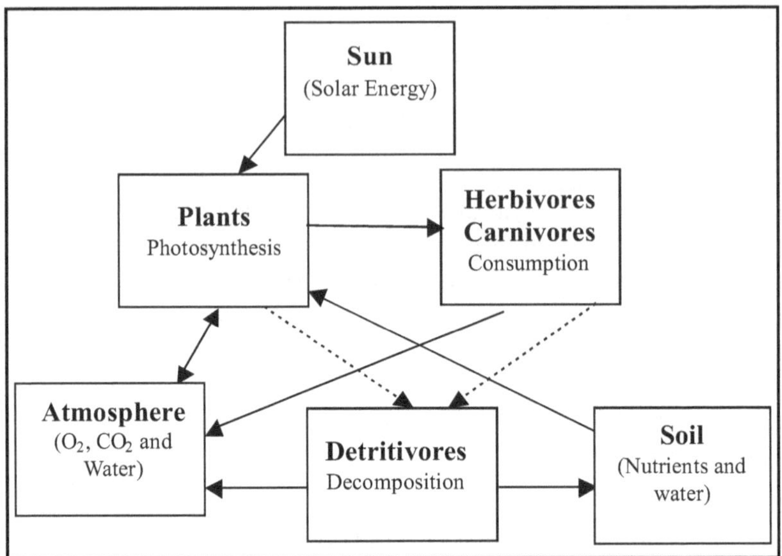

Fig. 2.1. Ecosystem model

proportion of the sunlight received at the Earth's surface say 1 to 4 % is transformed into biochemical form. Several studies have been carried out to determine this amount. Most ecosystems fix less than 2 % of the sunlight available for photosynthesis.

Living organisms can use energy in basically two forms called radiant and fixed:. Radiant energy exists in the form of electromagnetic energy, such as light. Fixed energy is the potential chemical energy found in organic substances. This energy can be released through respiration. Organisms that can take energy from inorganic sources and fix it into energy rich organic molecules are called autotrophs. If this energy comes from light then these organisms are called photosynthetic autotrophs. In most ecosystems plants are the dominant photosynthetic autotroph.

Organisms that require fixed energy found in organic molecules for their survival are called hetrotrophs. Heterotrophs who obtain their energy from living organisms are called consumer. Consumers can be of two basic types: Consumer and decomposers. Consumers that consume plants are know as herbivorous. Carnivorous are consumers who eat herbivores or other carnivores. Decomposer namely detrivores are heterotrophs that obtain their energy either from dead organisms or from organic compounds dispersed in the environment.

Organic energy can move within the ecosystem through the consumption of living or dead organic matter after fixation by plants. After decomposition the chemicals that were once organized into organic compounds are returned to their inorganic form and can be taken up by plants once again. Organic energy can also move from one ecosystem to another by a variety of processes like animal migration, animal harvesting, plant harvesting, plant dispersal of seeds, leaching, and erosion. Fig.2.2 shows the various inputs and outputs of Energy and Mater in a typical Ecosystem.

2.1.3 Various Types of Ecosystems

There are many different ecosystems: rain forests and tundra, coral reefs and ponds, grasslands and deserts. Climate differences from place to place largely determine the types of ecosystems; Let us see all types of Ecosystem.

(i) Mountains

Salient features

Mountains are a common sight on the earth with make up of one-fifth of the world's landscape, and provide homes to at least one-tenth of the world's people. About, 2.4 billion people rely on mountain ecosystems for most of their food, hydroelectricity, timber, and minerals. About 75-83 per cent of the earth fresh water originates from mountains. About half of the world's population relying on mountains for fresh water with the increasing population and standard of living, it is of utmost important to protect the mountain

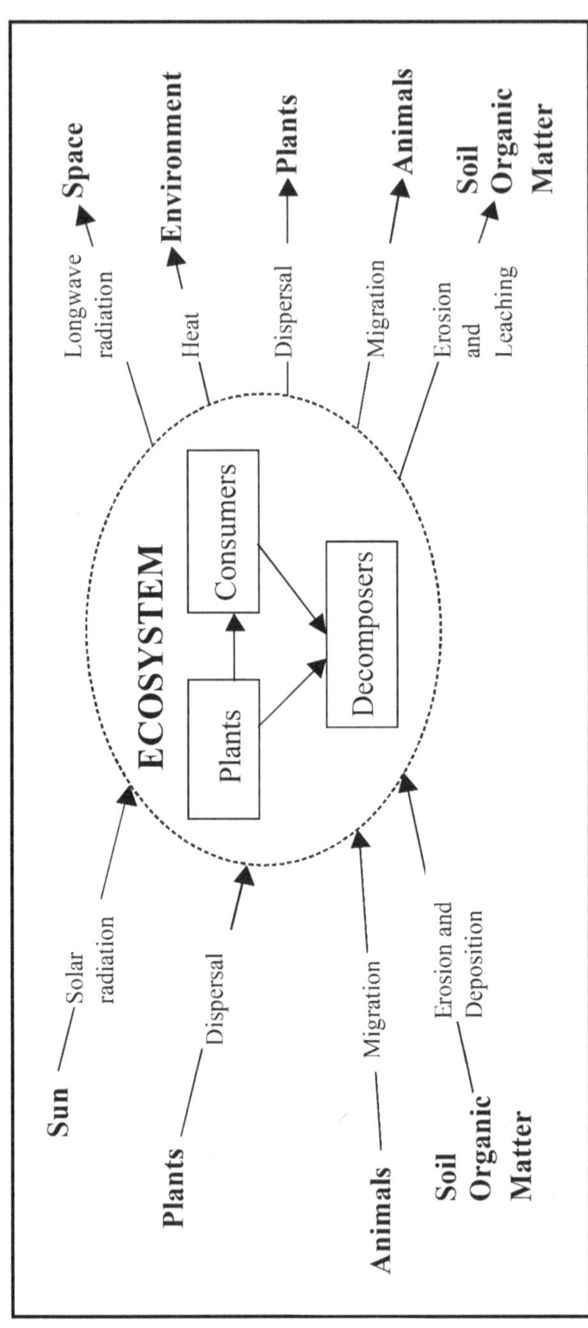

Fig 2.2. Energy and Matter Flow in a typical Ecosystems

biome. Generally mountain ecosystems have - very rapid changes in altitude, climate, soil, and vegetation over very short distances. Mountain provides home and shelter to many of our planet's ethnic minorities and animals and this ecosystems pose a high range of biodiversity due to changing human activity with the scientific advancements. As the more and more the indigenous people called tribal in mountain are being kicked out of their homes due to population and commercial growth, logging, and mining, one has to protect the culture from challenging environment. For example, the mountain's wide variety of organisms can be seen in California's Sierra Nevada range. It has been estimated that this range alone houses 10,000 to 15,000 different species of plants and animals. This is all mainly due to elevation changes, which produces belts, or zones, of differing climates, soils, and plant life.

Climatical conditions

Rainfall varies greatly all along the world's montane (mountain) biomes depending on geographical location, The rainfall varying from very wet to very dry and . swift weather changes in all the biomes. For example, in just a few minutes a thunder storm can roll in when the sky was perfectly clear, and in just a few hours the temperatures can drop from extremely hot temperatures to temperatures that are below freezing.

Human Activities

Mountains provide a home to several thousand different ethnic groups like of indigenous people, ethnic minorities, and refugees. The people in mountain have several occupation with classification as nomads, hunters, foragers, traders, small farmers, loggers, and miners, etc. These groups have been able to cope with this harsh environment of the mountain ecosystem. Most of the mountain people are in poverty with the lack in material wealth with which they make up in community life. This poverty is mainly due to widespread destruction and deforestation. Plants and animal species have to be preserved by people in Mountain. For example, in India's Garhwal Himalaya, local women were recently successful in identifying over 145 species of plants.

(ii) Tundra

Climatical conditions

Tundra is in the world in the highest northern latitude and in the southern hemisphere is found only in the Antarctic Peninsula and Islands close by.. Tundra region covers approximately one-fifth of the Earth's land surface. Here the temperatures often reach about – 50 to 70°F in the winter with cold climate which is impossible for trees to grow, thereby leaving room for low-growing plant life and wildflowers. For this reason the Tundra biome looks like a frozen-over prairie land.. In the Ice Age, massive glaciers dwelled here. As the Earth warmed these glaciers retreated leaving bare rock and scoured soils. The freezing temperatures leave deeper layers of soil frozen throughout most of the year. This freezing condition is called permafrost. Only the top layer on the surface is able to thaw out in summer conditions. However, this occurs just briefly because summers are extremely short here. The combination of a harsh climate, the lack of nutrients in the soil making it very hard for any type of plant life to grow.

Animals and their activity

Animals in this region still manage to survive here in spite of tough living conditions. During the summer, insects hatch out of eggs which were frozen in the top soil. This condition will acts as a vast feeding ground for thousands of migrating birds. Millions of migrating waterfowl and shore birds come to the shore and lake areas in the artic tundra of Alaska during the summer months. Ravens, hawks, ptarmigans, and the open country owl are also common migrating birds. Besides birds and insects mammals dwell in this icy zone. The Caribou, artic hare, mink, weasel, lemming, wolf, wolverine, brown bear, vole and reindeer, roam this land. And the *Polar bear*, walrus and artic fox are commonly seen on the ice pack and coastal areas.

(iii) The temperate forest

Temperate forests are located in the mid-latitudes in eastern North America, Western Europe, and eastern Asia. In the southern hemisphere, smaller areas of temperate forest can be found in South America, southern Africa, Australia, and New Zealand. Climate in temperate forests region are highly seasonal with the warm summers and cold winters. The trees change colors as the seasons cycle: the green leaves of summer give way to the grey bare branches of winter. Temperate forests blend into the pines and firs of the taiga also called clod climate forest. The temperate forest biome is one of the fast changing biomes on the earth due to reason that population density very closely corresponds to the distribution of temperate forests as this type of forest is providing good potential for various occupation. Trees in temperate forest used for construction, firewood and art leading to the decline and loss of these forests in many parts of the world.

(iv) The Desert region

The desert biome has the driest climate compared with the other biomes in the globe. The temperatures lie between 20 degrees to 30 degrees north and south latitude. It is here that equatorial air falls down toward the Earth's surface and rainfall is rare because the equatorial air that is falling prevents most air from rising. The globe region covered in the major desert are North Africa, southwestern North America, the Middle East, and Australia. In addition there are smaller deserts such as on the Pacific coast of South America (the Atacama) and the Atlantic coast of southern Africa (the Namib). In these small desert area moisture from cold water currents is evaporated immediately by the hot land masses adjacent to the currents.

Salient feature of desert

Rainfall in the desert region is usually a few inches yearly or, in some regions absolutely no rain. Desert soils are often salty due to quick evaporation of even scarce rain water from the ground. Due to scarcity of water, plants in the desert are almost always drought-tolerant, meaning they can survive without water for a long time with the unique features such as, thick or waxy leaves, large root systems, and water storage systems-like in the cactus. These kinds of adaptive nature plants have ability to store water, find water quickly or live with the littlest amount of water possible. The vegetation in deserts varies very widely. For example, the molar desert in California is known for it's unique Joshua trees, whereas in the Sonoran desert there is the thorn-covered Ocotillo and the giant Saguaro Cactus. Sagebrush types are found in Great Basin in the western United States.

Animal activities

The animals live in the desert are usually light-colored and use camouflage to blend into their surroundings, and possibly for protection against predators. Animals are more active at night and around dawn and dusk and during the day they often lay in burrows or under rocks to escape the scorching heat. Different species of life in the desert region are jackrabbits (North America), kangaroo rats, owls, snakes, lizards and tortoises.

(v) Tropical Dry Forests

Tropical dry forests have high temperatures throughout the entire year. These are found primarily in Central

America, southern Asia and some parts of Australia The forest in south Asia is called Monsoon Forests. Climate hers is a clear dry season which limits plant growth and the activity of animals. There is really no clear-cut distinction between this zone and the tropical rain forest because the length of the dry season varies tremendously throughout the tropics.

(vi) Cold Climate Forests

This cold climate is found at very high latitudes extending across Eurasia and North America. Rainfall in this climate is moderately high all throughout the year. In the winter snow is covering the ground with the little evaporation of water by the sun. Usually ponds, lakes and bogs also known as "muskegs" are found everywhere, especially in glacially carved areas. Coniferous trees are found in this region.

Vegetation

Trees in the cold climate forest use a lot of energy to grow their leaves, thus they have found a way to keep their needles through out the year. This way, when the sun comes out again in the spring these trees are already gathering much needed sunlight instead of wasting more energy to grow new leaves. In addition they have adapted their needles to be filled with a chemical that repels grazing animals, and their thick bark resists the loss of moisture in the cold winters. Trees in this region known as boreal or the Northern coniferous forests, usually have shrubs underneath them with blueberries (which is a favorite food of many animals) which act as heath plants. All along the river banks throughout the taiga, willows and many other well known trees can be found. Leaves cover the ground for the relatively low temperature and the acidic soil slows down the process of decay.

Animals

Many animals migrate to the cold climate forest in the summer months. However, those who do not have learned to adapt to the cold. Moose, wolves, woodland caribou, wood bison, black bear, marten, lynx, and the arctic ground squirrel are the common animals found in this region. Birds species, such as the redpoll, raven, gray jay, red-throated loon, northern shrike, sharp-tailed grouse, and fox sparrow are found only in summer. Also the bald eagle, peregrine falcon, and osprey which are fish eaters, live in this biome.

Human Activities

Fishing, hunting and trapping are common activities that take place in among the people living in this region. In addition, tourism, mining, oil and gas extractions and forestry are main activities which occur in this biome. Many of these human activities affect the natural systems in this ecosystem.

(vii) The Grasslands

Climatical conditions

Grasslands stretch basically thousands of miles primarily in the continents of North America and Asia. A limited area of grasslands lies across southern South America. Two extremes of climate namely hot summer and cold winter is prevailing in this region. In hot summers, the flora is getting baked into a nice crisp golden brown, and in the winter the land freezes over. There is a powerful winds with high speed in this region because of the vast open area. The seasonal changes always have helped maintain this biome. However not just the seasonal changes can keep this biome in check. Wildfires help maintain this biome.

Vegetation

Rainfall in the grasslands is moderate, but the e amount of rain is not sufficient enough to support clusters of trees.

Regions where more rainfall falls than another area results in taller grasses that can reach impressive heights. These grasses dominate the vegetation in this biome. Besides the grasses, wildflowers every spring provide a beautiful display of color of green rainbow-filled landscape.

Human Activities: The Grasslands soil is usually deep, dark, and rich, called mollisols, in drier regions it is called aridisols. Because a lot of its soil is very rich, a lot of the grasslands have been destroyed and most of it disturbed, due to farming. Grasslands are now major regions for growing crops, like wheat, corn, and other grains.

(viii) Tropical Grasslands

Climatical conditions

This is also called as savannahs and are found in entire continent of Africa, little- in India and the northern part of South America. Savannahs are located at tropical latitudes, and have however much dry climate than many tropical forests. Rainfall in this biome is between 20 to 60 inches a year with the seasonal change of once in two weeks.

Vegetation

Throughout the savannahs, one can see plants like grasses and small plants. Trees are growing in the area where there are cracks in the surface or deep soil. In many savannahs around the world palm trees play an important role in the landscape. The most dominate wooded form in the savanna are the thorn woodlands. There is a large amount of wild fruit-trees, which provide food for many birds and animals.

Animals

Birds like shrikes, hornbills, grey louries, flycatchers, knysna, purple-crested louries, green pigeons, rollers and raptors are found in this region. Mammals like lions, leopards, cheetahs, elephants, buffalos, rhinoceroses, giraffes, hippopotami, gazelles, zebras, kudus, waterbucks, oryxes are found in Savannahs.

(ix) Tropical Rainforests

Tropical rainforests are one of major source of treasures for the world as the tropical rainforests are a critical link in the ecological chains of our earth's biosphere, and most of them are now at risk because of increasing human activity. Already more than half of the world's original tropical rainforests was destroyed. If this trend persists in a few decades, all rainforest get eliminated. Leading to major ecological imbalance and so the tropical rainforest have to be preserved.

Forest in the tropics are receiving 4-8 meters of rain each year and are found in Central and South America, Southeast Asia and islands near it, and West Africa. There are smaller rainforests in northern Australia and other small islands. All tropical rainforests are found along the equator where the temperatures and the humidity are always high, with the equal time period of day and night.

Generally, there is a flow of air that comes from the poles of the Earth, towards the equator. These winds are filled with moisture and the intense heat that is located at the equator causes the moisture to rise, cool and then condense to create rain. This continuous cycle causes it to rain almost 24 hours a day around most of the tropical areas.

(x) Aquatic Ecosystem

An aquatic ecosystem is an ecosystem that is based in water, whether it is a pond, lake, river, underground water body, estuary or ocean. It consist of living aquatic organisms (eg: fish, plankton, annelids, etc.) which constitute the biota of the ecosystem, as well as their physical environment which collectively can be referred to as the biotope.

Aquatic ecosystems are generally classified into two types namely Inland and Marine

1) Inland aquatic, or fresh water 'ecosystems: This ecosystems may be a. running water (spring, stream) or standing water (pond, puddles). Subterranean ecosystems exist in many localities, and are particularly common in karst landscapes. Such ecosystems are at present widely neglected by established systems of protected areas.
2) Marine (salt water) ecosystems: These ecosystems may include both shallow coastal waters and the deep waters of the open ocean.

2.2 Biodiversity

Biodiversity or biological diversity is the sum total of the variety of life and its interactions and can be subdivided into 1) Genetic Diversity 2) Species Diversity, and 3) Ecological or Ecosystem Diversity.

2.2.1 Genetic Diversity

Genetic diversity is the degree of variability of the genetic material of an organism. Species are defined by the differences in their genes. High genetic diversity indicates populations that can more easily adapt to changing situations and environments, and also a greater assortment of materials that can be found, increasing the chances of finding a useful compound.

However, exact assessment of genetic diversity is both time-consuming and prohibitively expensive, requiring modern laboratories and expensive chemicals.

2.2.2 Species Diversity

Fortunately, genetic diversity can be estimated by species diversity, and this has become the standard unit of measurement in most biodiversity surveys

Species have the advantage of being natural biological divisions and easily identifiable; their diverging appearances were the basis by which they were classified in the 18th century, and modern phylogenetic techniques more often than not produce species divisions similar to those of classical taxonomic divisions. For many groups of organisms, such as birds and flowers, public interest means that identification of many species is already known by large numbers of people.

The degree of genetic variability at the species level, and indeed at any taxonomic level, can be maximized by taking species that differ by one another by as many characters as possible. If these characters represent different genetic elements, then the divergent species should represent greater genetic diversity.

2.2.3 Ecosystem Biodiversity

Ecosystems have six main parts that make up or affect them. They are the sun, abiotic substances (non-living things such as water or climate), primary producers, primary consumers, secondary consumers, and decomposers. The sun and abiotic substances provide the energy for primary producers to grow. The sun provides sunlight, and abiotic substances provide nutrients such as water. The primary consumers are the animals that eat the plants (the primary producers). The predators - the animals that eat other animals - are the secondary consumers. Bacteria and fungi are the decomposers that break down dead plants and animals to become nutrients in the soil.

(i) Effects of Biodiversity

Changes occur all of the time in ecosystems. The changes occur daily, seasonally, or over many years. Ecosystems tend to remain in balance unless altered by humans or natural events such as hurricanes, floods, or droughts. Sometimes people impact many ecosystems in order to obtain a desired product or satisfy basic needs for food, shelter, and clothing. The habitats of any ecosystem can be protected by responsible human decisions about how we live, produce food, and use energy. Ecologists are concerned about the rate at which nonrenewable resources such as coal, gas, and petroleum are being depleted from our environment. However, many new uses for soybeans and corn in the agricultural industry can help these problems.

(ii) Some remedies for Biodiversity

Some uses of soybeans are biodiesel fuel, soy crayons, and soy ink. These three products are made with soybean oil instead of petroleum. Biodiesel fuel is then mixed with petroleum diesel to reduce the amount of petroleum diesel used. By using soybean oil instead of petroleum, we decrease our dependence on petroleum, which is a nonrenewable resource.

Some uses for corn are biodegradable plastics, packaging material, and an alternative fuel called ethanol. Biodegradable plastics and packaging materials (also known as packing peanuts) are made with corn starch and easily break down so they are not harmful to the environment. Ethanol is a fuel made with corn oil. By using an ethanol-blend fuel, hazardous carbon monoxide emissions are reduced by 25-30 percent.

2.2.4 Values of Biodiversity

Many conclusions can be made from scientific, philosophical, economic, ethical, and aesthetic perspectives. Scientists argue that much remains to be learned about many species and ecosystems around the world and that the loss of these species would foreclose that opportunity. For example loss of certain rare species like sumatran tiger and Rhinoceros are worth to be preserved. There is great beauty in forests, coral reefs, savannahs, and other landscapes that is worth preserving for future generations, as well as our own, to appreciate.

(i) Instrumental Values

Materialistic uses of biodiversity are the core of instrumental values and are as follows

- Provision of food, fuel and fibber
- Provision of shelter and building materials
- Purification of air and water
- Detoxification and decomposition of wastes
- Stabilization and moderation of the Earth's climate
- Moderation of floods, droughts, temperature extremes and the forces of wind
- Generation and renewal of soil fertility, including nutrient cycling
- Pollination of plants, including many crops
- Control of pests and diseases
- Maintenance of genetic resources as key inputs to crop varieties and
- Livestock breeds, medicines, and other products
- Ability to adapt to change

(ii) Spiritual Values

Sometimes, our feeling for biodiversity are revealed in the way we spend our money; sometimes they are not. Economists are trying to devise methods for estimating the monetary value of the blue whales and harp seals for people whose only relationship with them is knowing that they exist.

(iii) Ecological Values

Every population of every species is part of an ecosystem of interacting populations and environment and thus has an ecological role to play. There are products, consumers, decomposers, and many variations of these roles and others- competitors, dispersers and pollinators, and more. Some species play ecological roles that are of great importance than we would predict form their abundance; these are called keystone species. However, when assessing the ecological roles of species, conservation biologists are typically conservative and assume every component of an ecosystem is critical until less proven otherwise. A species that is relatively unimportant now may become important as an ecosystem changes through time. For example, during the last 12,000 years the eastern white pine has varied form being quite rare to being an ecosystem dominant over large area.

With a large agenda and limited resources, conservation biologists have to be efficient strategists, and this often leads to them to target certain species to advance their overall goal of maintaining biodiversity. Best known is the flagship species. The charismatic species that have captured the public's heart and won their support for conservation. Some species have won concert to conservation across the globe; consider the cuddliness of the giant panda, haunting songs of humpback whale, and the grandeur of the tiger or gorilla. Those species are of important conservation strategic values.

(iv) Potential Values

Potential values generally focus on their usefulness hear and now, but this is a shortsighted viewpoint as revealed in our discussion of medicinal research and biodiversity. Our rudimentary understanding of biology and ecology have leave an enormous gap between the currently realized values of a species and its potential future value. This gap is particularly wide because we have vague idea of what our future lives will be like- technologically, culturally and ecologically. It may be difficult to guess at the potential ecological role that a species might assume in the future. It would certainly have taken a very prescient biologists to guess that the shrewlike mammals tat shared the earth with dinosaurs would lead to the earth-dominating *Homo sapiens*. The evolution of our scientific knowledge has already taught us that: keep options alive. We must take this approach because we know so little. We can never say of species, a part of biodiversity that lacks of values.

(v) Scientific and Educational Values

The world is a complex place, but our knowledge of it is increasing all the time. And some of the credit goes to our fellow inhabitants. Of cause scientific inquiry is just an advanced form of the intellectual curiosity about the world that begins in infancy. Our education would suffer greatly without a diverse world to explore, without the bean seeds to plan, without the frog eggs to watch develop into tadpoles. Whether we want to learn about ourselves or the world we share with other species, we need models to observe.

(vi) Intrinsic Values

Environmental philosophers who claim that intrinsic value exist objectively in human and other organisms. All organisms strive (usually unconsciously and in an evolutionary sense) to achieve certain basic predetermined goals- to grow, to reach maturity and to reproduce. Intrinsic value of biodiversity is non-anthropocentric. There is a general opinion that recognizing the intrinsic value of biodiversity has a dramatic effect upon the framework of environmental debate and decision making.

2.2.5 Threats of biodiversity

Due to human actions, species and ecosystems are threatened with destruction to an extent rarely seen in Earth history. The human action is interpreted in two ways. First one is the loss of species and ecosystems to the accelerating transformation of the Earth by a growing human population. It is estimated that as the human population passes the 6 billion mark, half of the word's forests would be destroyed. We appropriate roughly half of the world's net primary productivity for human use. Secondly there are six specific types of human actions that threaten species and ecosystems. These are as follows.

(i) Over-Hunting

Over-hunting has been a significant cause of the extinction of hundreds of species and the endangerment of many more, such as whales and many African large mammals. Most extinctions over the past several hundred years are mainly due to over-harvesting for food, fashion, and profit. Commercial hunting, both legal and illegal (poaching), is the principal threat.

The pet and decorative plant trade falls within this commercial hunting category, and includes a mix of legal and illegal activities. The annual trade is estimated to be at least US $5 billion, with perhaps one-quarter to one-third of it illegal. Sport or recreational hunting causes no endangerment of species where it is well regulated, and may help to bring back a species from the edge of extinction. Many wildlife managers view sport hunting as the principal basis for protection of wildlife.

(ii) Habitat Loss, Degradation, Fragmentation

Habitat loss, degradation, and fragmentation are important causes of known extinctions. As deforestation proceeds in tropical forests, this promises to become the main cause of mass extinctions caused by human activity.

All species have specific food and habitat needs. The more specific these needs and localized the habitat, the greater the vulnerability of species to loss of habitat to agricultural land, livestock, roads and cities. In the future, the only species that survive are likely to be those whose habitats are highly protected, or whose habitat corresponds to the degraded state associated with human activity (human commensals).

Habitat damage, especially the conversion of forested land to agriculture (and, often, subsequent abandonment as marginal land), has a long human history. It began in China about 4,000 years ago, was largely completed in Europe by about 400 years ago, and swept across the US over the past 200 years or so. Viewed in this historical context, we are now mopping up the last forests of the Pacific Northwest

Tropical forests are so important because they harbor at least 50 percent, and perhaps more, of the world's biodiversity. Direct observations, reinforced by satellite data, document that these forests are declining. The original extent of tropical rain forests was 15 million square km. Now there remains about 7.5-8 million square km, so half is gone. The current rate of loss is estimated at near 2 percent annually (100,000 square km destroyed, another 100,000 square km degraded). While there is uncertainty regarding the rate of loss, and what it will be in future, the likelihood is that tropical forests will be reduced to 10-25 percent of their original extent by late twenty-first century

Habitat fragmentation is a further aspect of habitat loss that often goes unrecognized. The forest, meadow, or other habitat that remains generally is in small, isolated bits rather than in large, intact units. Each is a tiny island that can at best maintain a very small population. Environmental fluctuations, disease, and other chance factors make such small isolates highly vulnerable to extinction. Any species that requires a large home range, such as a grizzly bear, will not survive if the area is too small. Finally, we know that small land units are strongly affected by their surroundings, in terms of climate, dispersing species, etc. As a consequence, the ecology of a small isolate may differ from that of a similar ecosystem on a larger scale.

For the future, habitat loss, degradation, and fragmentation combined is the single most important factor in the projected extinction crisis.

(iii) Invasion of non-native Species

Invasion of non-native species is an important and often overlooked cause of extinctions. The African Great Lakes—Victoria, Malawi and Tanganyika—are famous for their great diversity of endemic species, termed "species flocks," of cichlid fishes. In Lake Victoria, a single, exotic species, the Nile Perch, has become established and may cause the extinction of most of the native species, by simply eating them all. It was a purposeful introduction for subsistence and sports fishing, and a great disaster.

Of all documented extinctions since 1600, introduced species appear to have played a role in at least half. The clue is the disproportionate number of species lost from islands: some 93 percent of 30 documented extinctions of species and sub-species of amphibians and reptiles, 93 percent of 176 species and sub-species of land and freshwater birds, but only 27 percent of 114 species and subspecies of mammals. Why are island species so vulnerable, and why is this evidence of the role of non-indigenous species? Islands are laboratories for evolution

(iv) Domino Effects

Domino effects occur when the removal of one species (an extinction event) or the addition of one species affects the entire biological system. Domino effects are especially likely when two or more species are highly interdependent, or when the affected species is a "keystone" species, meaning that it has strong connections to many other species.

A keystone species is one whose influence on others is disproportionately great. A seminal study of marine invertebrates in the rocky intertribal region of Washington State found that the top predator, a starfish, facilitated the coexistence of many other invertebrates by selectively consuming mussels, which otherwise would crowd out other organisms. Thus a keystone species is one whose presence or absence both directly and indirectly influences other species through food web connectivity. Contrary to what some may think, not all species are "keystones", and it requires careful experimental studies to identify keystone species.

(v) Pollution

Pollution from chemical contaminants certainly poses a further threat to species and ecosystems. While not commonly a cause of extinction, it likely can be for species whose range is extremely small, and threatened by contamination. Several species of desert pupfish, occurring in small isolated pools in the US Southwest, are examples

(vi) Climate Change

A changing global climate threatens species and ecosystems. The distribution of species (biogeography) is largely determined by climate, as is the distribution of ecosystems and plant vegetation zones (biomes). Climate change may simply shift these distributions but, for a number of reasons, plants and animals may not be able to adjust. The pace of climate change almost certainly will be more rapid than most plants are able to migrate.

The presence of roads, cities, and other barriers associated with human presence may provide no opportunity for distributional shifts. Parks and nature reserves are fixed locations. The climate that characterizes present-day Yellowstone Park will shift several hundred miles northward. The park itself is a fixed location. For these reasons, some species and ecosystems are likely to be eliminated by climate change. Mountaintop species are especially vulnerable. The plants and animals found on high mountains of the American West include many remnants of a Pleistocene fauna that long ago was displaced toward the arctic, or upslope. With further warming, many of these mountaintop species likely will be eliminated

A changing climate will have many other effects. The southern extent of the Everglades, today the site of the most ambitious and expensive restoration project ever undertaken, may be underwater, along with significant areas of human habitation. Agricultural production likely will show regional variation in gains and losses, depending upon crops and climate. Some coral reefs will expand, and others will contract or die off. Ecological changes due to an altered climate are difficult to forecast, but expected to be serious.

As a consequence of these multiple forces, many scientists fear that by end of next century, perhaps 25 percent of existing species will be lost.

2.2.6 Endangered and Endemica species of India

India has a total of 89,451 animal species accounting for around 8 % of the faunal species in the world and the flora accounts for 11 % of the global total. The endemism of Indian biodiversity is high - about 33-35 % of the country's recorded flora. These endemic are concentrated mainly in the North-East, Western Ghats, North-West Himalayas and the Andaman and Nicobar islands. The main factors contributing to this bio-diversity are due to habitat destruction, degradation, fragmentation and over-exploitation of resources.

It is reported that 44 plant species are critically endangered, 113 endangered and 87 vulnerable. as far as plant species are concerned. Among the animals, 18 are critically endangered, 54 endangered and 143 are vulnerable. India ranks second in terms of the number of threatened mammals and ranks sixth in terms of countries with the most threatened birds.

Typical data of Threatened Species of India by taxonomic group

Taxonomic group	Number of threatened species
Mammals	84
Birds	72
Reptiles	24
Amphibians	3
Fish	4
Molluscs	2
Other Invertebrates	22
Plants	246
Total	**457**

2.2.7 Biodiversity Conservation

Conservation is the protection, preservation, management, or restoration of wildlife and natural resources such as forests and water. Through the conservation of biodiversity the survival of many species and habitats which are threatened due to human activities can be ensured. Other reasons for conserving biodiversity include securing valuable Natural Resources for future generations and protecting the well being of eco-system function

Conservation can broadly be divided into two types:

In-situ: Conservation of habitats, species and ecosystems where they naturally occur. This is in-situ conservation and the natural processes and interaction are conserved as well as the elements of biodiversity.

Ex-situ: The conservation of elements of biodiversity out of the context of their natural habitats is referred to as ex-situ conservation. Zoos, botanical gardens and seed banks are all example of ex-situ conservation.

In-situ conservation is not always possible as habitats may have been degraded and there may be competition for land which means species need to be removed from the area to save them.

(i) In situ Conservation Methods

In-situ conservation of biodiversity is quite simply the maintenance of species in their natural habitats

In the case of cultivated species, *in situ* conservation is the conservation of species in the surroundings in which they developed their distinctive properties.

The aim of *in situ* conservation is to enable biodiversity to maintain itself within the context of the ecosystem in which it is found. In the case of plants, it allows them to evolve and develop as part of the ecosystem of their natural habitat.

On-farm conservation is one form of *in situ* conservation. Farmers have long been been aware of the relationship between the stability and sustainability of their production systems and the diversity of the crops they grow. In fact, farmers created most of the crop diversity we see today, by selecting the various qualities they wanted from each year's harvest.

Home gardens have also recently been recognized as central to *in situ* conservation. They often serve as refuges for crop and crop varieties that were once more widespread in the larger agro-ecosystems. Farmers use home gardens as a space for plants with little or no world market value. This may be because they offer important nutrient combinations or are of special importance to local food culture. They may also hold religious significance for the farmer. Home gardens often also serve as sites for experimentation and introduction of new cultivars. It is therefore crucial to understand their dynamics so that they can take their place as a component of *in situ* conservation of global agro-biodiversity.

Nature parks and other protected areas have long been important centers for *in situ* conservation. Around 400 botanic gardens worldwide are active in this kind of conservation. Several botanic gardens either maintain or manage natural reserves and areas of natural vegetation and work closely with managers of national parks and other protected areas.

(ii) Ex-situ Conservation

The first step is to select hotspots of diversity. Since it is not possible to conserve all biodiversity due to lack of resources and the need to use land for human activities, areas are prioritized to those which are most in need of conservation. 'Hotspot' a term used to define regions of high conservation priority combining high richness, high endemism and high threat.

Test Your Understanding

1. What is Ecosystem? Explain
2. Discuss the basic concepts of Ecosystem
3. Explain Energy and Matter flow in the Ecosystem
4. Discuss the Ecosystem of Tundra region
5. Discuss the Ecosystem of Tropical grassland
6. What is Biodiversity? Explain
7. Narrate the potential values of biodiversity
8. Explain various threats of Biodiversity
9. Discuss the Biodiversity conservation

CHAPTER 3

Environmental Pollution

3.1 Air Pollution

Air pollution is a phenomenon by which particles either a solid or liquid and gases contaminate the environment. Such contamination can result in health effects on the population, which might be either chronic arising especially from long-term exposure or acute due to accidents. Other effects of pollution may be damage to materials, agricultural damage namely reduced crop yields and tree growth, impairment of visibility namely tiny particles scatter light very efficiently, and even climate change global warming.

3.1.1 Classification of Air Pollutants

Air Pollutants are classified as Natural as well as Anthropogenic. *Natural pollutants* are those that are found in nature or are emitted from natural sources. For example, volcanic activity produces sulfur dioxide, and particulate pollution may derive from forest fires or windblown dust. *Anthropogenic pollutants* are those that are produced by humans or controlled processes. For example, sulfur dioxide is produced by fossil fuel combustion and particulate matter comes from diesel engines.

Air pollutants are further classified as *primary* or *secondary*. Primary pollutants are those that are emitted directly into the atmosphere from an identifiable source. Examples include carbon monoxide and sulfur dioxide. Secondary pollutants are those that are produced in the atmosphere by chemical and physical processes from primary pollutants and natural constituents. For example, ozone is produced by hydrocarbons and oxides of nitrogen (both of which may be produced by car emissions) and sunlight.

3.1.2 Properties of Air Pollutants

Air pollutants are divided into two categories as Criteria Air Pollutants and Hazardous Air Pollutants. Criteria Air Pollutants are the primary ingredients of the most easily observed air pollution - urban smog and soot. They are the 6 most common air pollutants: Carbon Monoxide; Lead; Nitrogen Dioxide; Ozone (formed from precursor Volatile Organic Compounds); Particulate Matter; and Sulfur Dioxide. Particulate matters are available in six forms

Dust: It consists of particle of the size from 1 to 200 micron and formed by natural disintegration of rock and soil or by the mechanical process of grinding and spraying.

Smoke: It is the fine particles of size ranging from 0.01 to 1 micron which can be liquid or solid and formed by combustion or other chemical process. The color also will vary depending on the nature of material burnt

Fumes: Very fine solid released from chemical or metallurgical process with the size between 0.1 to 1 micron

Mist: Fine droplets formed by condensation in the atmosphere or released from industries with the size less than 10 micron

Fog: It is mist in which liquid is water

Aerosol: It is all air-borne suspensions either solid or liquid, and the size smaller than 1 micron.

All the above particulate matter can be emitted directly or form in the atmosphere. "Primary" particles, such as dust from roads or elemental carbon (soot) from wood combustion, are emitted directly into the atmosphere. "Secondary" particles are formed in the atmosphere from primary gaseous emissions. Examples include sulfates, formed from SO_2 emissions from power plants and industrial facilities, and nitrates, formed from NO_x emissions from power plants, automobiles, and other types of combustion sources. The chemical composition of particles depends on location, time of year, and weather. Generally, coarse Particulate Matter is composed largely of primary particles and fine particulate matter contains many more secondary particles.

Hazardous Air Pollutants (HAPS) - HAPs are the air-borne chemicals that can cause adverse effects to human health or the environment. Almost 200 of these chemicals have been identified, including chemicals that can cause cancer or birth defects.

3.1.3 Health Effects

Particulate Pollutants: Fine particles are the major cause of reduced visibility in parts of the world. Also, soils, plants, water, or materials are affected by particulate matter. For example, particles containing nitrogen and sulfur that are deposited as acid rain on land or water bodies may alter the nutrient balance and acidity of those environments so that species composition and buffering capacity change. Particulate matter causes soiling and erosion damage to materials, including culturally important objects such as carved monuments and statues.

Carbon Monoxide: Carbon monoxide (CO) is a colorless, odorless, and at high levels a poisonous gas that is fairly unreactive. It is formed when carbon in fuels is not burned completely. The major source of CO is motor vehicle exhaust. In cities, as much as 95 percent of all CO emissions result from vehicular (automobile) emissions. Other sources of CO emissions include industrial processes, nontransportation-related fuel combustion, and natural sources such as wildfires.

CO has serious health effects on humans: An exposure to 50 ppm of CO for eight hours can cause reduced psychomotor performance, while CO is lethal to humans when concentrations exceed approximately 750 ppm. Hemoglobin, the part of blood that carries oxygen to body parts, has an affinity of CO that is about 240 times higher than that for oxygen, forming carboxyhemoglobin, COHb. Moreover, the release of oxygen by hemoglobin is reduced in the presence of COHb. However, the effects of CO poisoning are reversible once the CO source has been removed.

Sulfur Dioxide: Sulfur dioxide (SO_2) is colorless, nonflammable, non-explosive gas. Almost 90 percent of anthropogenic SO_2 emissions are the result of fossil fuel combustion (mostly coal) in power plants and other stationary sources. A natural source of sulfur oxides is volcanic activities.

In general, exposure to SO_2 irritates the human upper respiratory tract. The most serious air pollution episodes occurred when there was a synergistic effect of SO_2 with PM and water vapor (fog). Because of this, it has proven difficult to isolate the effects of SO_2 alone.

SO_2 is one of the precursors of acid rain (the term used to describe the deposition of acidic substances from the atmosphere). Also, SO_2 is the precursor of secondary fine sulfate particles, which in turn affect human health and reduce visibility. Prolonged exposure to SO_2 and sulfate PM causes serious damage to materials such as marble, limestone, and mortar. The carbonates (e.g., limestone, $CaCO_3$) in these materials are replaced by sulfates (e.g., gypsum, $CaSO_4$) that are water-soluble and may be washed away easily by rain. This results in an eroded surface.

Nitrogen Dioxide: Nitrogen dioxide (NO_2) is a reddish-brown gas. It is a lung irritant and is present in the highest concentrations among other oxides of nitrogen in ambient air. Nitric oxide (NO) and NO_2 are collectively known as NO_x.

Anthropogenic emissions of NO_x come from high-temperature combustion processes, such as those occurring in automobiles and power plants. Natural sources of NO_2 are lightning and various biological processes in soil. The oxides of nitrogen, much like sulfur dioxide, are precursors of acid rain and visibility-reducing fine nitrate particles.

Ozone: Ozone (O_3) is a secondary pollutant and is formed in the atmosphere by the reaction of molecular oxygen, O_2, and atomic oxygen, O, which comes from the photochemical decomposition of NO_2. Volatile organic compounds or VOCs (e.g., what one smells when refuelling the car) must also be present if O_3 is to accumulate in the atmosphere.

O_3 occurs naturally in the stratosphere and provides a protective layer from the sun's ultraviolet rays high above the earth. However, at ground level, O_3 is a lung and eye irritant and can cause asthma attacks, especially in young children or other susceptible individuals. O_3, being a powerful oxidant, also attacks materials and has been found to cause reduced crop yields and stunt tree growth.

Lead: The major sources of lead (Pb) in the atmosphere in the United States are industrial processes from metals smelters. Thirty years ago, the major emissions of Pb resulted from cars burning leaded gasoline. In 2002 only aviation fuels contain relatively large amounts of Pb. Pb is a toxic metal and can accumulate in the blood, bones, and soft tissues. Even low exposure to Pb can cause mental retardation in children.

Hazardous Air Pollutants

Also known as toxic air pollutants or air toxics are pollutants that can cause cancer and other serious health effects such as birth defects. An example of an air pollutant found in gasoline is benzene, toluene, and xylenes.

Volatile Organic Compounds (VOCs)

VOC's have properties of being a gas at room temperatures. Often 10 times higher concentrations indoors than outdoors, these air pollutants can have short and long term effects. These products all contain VOC's:

- Paints
- Lacquers
- Paint strippers
- Cleaning supplies
- Pesticides
- Building materials
- Office equipment
- Glues and adhesives
- Permanent markers
- Photographic solutions

Many of the cleaning agents also contain organic solvents. Eye, nose, and mouth irritation, headaches, dizziness, fatigue, allergic skin reactions can all be symptoms of inhalation of these chemicals. If these chemicals are used, there is plenty of ventilation and put the caps back on tight when done. Paint strippers, adhesive removers, and aerosol spray paints all contain the solvent methylene chloride, which the body can convert to carbon monoxide, so use extreme care when using these chemicals.

Benzene

Benzene is a known human carcinogen, so keep exposure to that at a minimum. Benzene is found in tobacco smoke, stored fuels, and paint supplies.

Formaldehyde

A major source of formaldehyde is building materials. Carpets, insulation foam, and particleboard all contain formaldehyde. Mobile homes and new homes with pressed-wood materials can contain significant amounts of formaldehyde. Headaches, dizziness, nausea, and other eye, respiratory, and skin irritations are all common symptoms. Environmental experts disagree as to what is a safe limit of formaldehyde for the general public.

Radon

Indoor radon is the second leading cause of lung cancer among non-smokers. Radon is a colorless, odorless, and tasteless natural radioactive gas released from the earth. This air pollutant enters the environment through the soil, through uranium and phosphate mines, and through coal combustion. Because it is a heavy gas, it tends to collect in basements, entering through spaces in the soil or fill material around a home's foundation. Detectors are available that will measure the amount of radon in your home.

Perchloroethylene

This chemical is used most widely in dry cleaning. When the clothes come back from the dry cleaners, make sure there is no chemical smell emanating from them. At higher concentrations, perchloroethylene can range from dizziness and headaches to excessive sweating and unconsciousness.

3.1.4 Particulate Pollution Control Equipments

Particulate pollution control equipments working on the principles gravitational settling, centrifugal and inertial impaction, direct interception, diffusion, electrostatic charging and precipitation with the name of Gravitational settling chamber, Cyclone separator, Fabric filter, Electrostatic precipitator and Wet scrubber.

(i) Gravitational Settling Chamber

Particles of size more than 60 micron may be removed by the principles of gravitational settling. It consists of rectangular or square chamber of huge size. The dust laden gas enters at low velocity say less than 0.5 m/s depending on the particle size and density and the enough time will be allowed to settle the particle by gravity obeying the strokes law of settling. The equipment offers low maintenance and pressure drop. The efficiency of separation depends of particle size and density. As density and particle size decreases efficiency decreases. Gravity tray type settling chamber is also available with fixed number and variable number of trays to suit the required capacity.

(ii) Cyclone Separator

In this equipment centrifugal force is utilized for separating particle. This device consists of two chambers namely cylindrical and conical chamber. The cylindrical chamber is attached with tangential inlet for aiding circulator motion for the dust laden gas. The moment, the

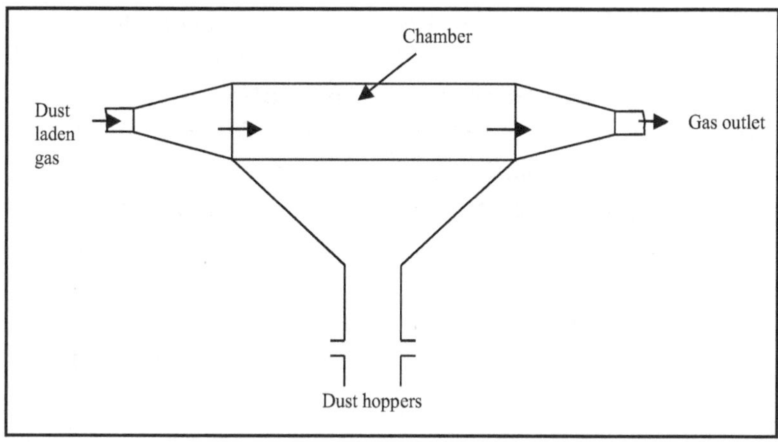

Fig. 3.1 Gravity settling chamber

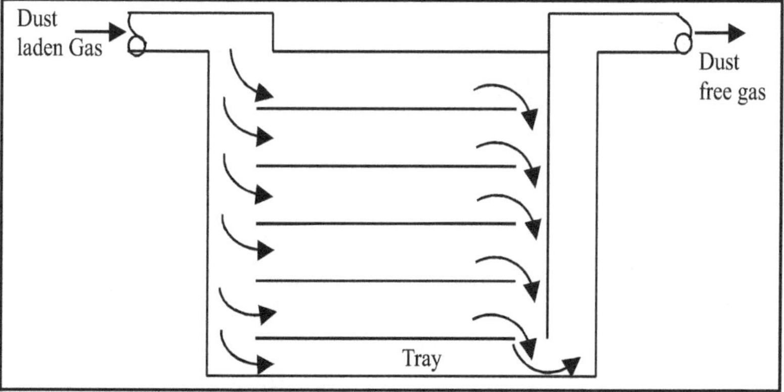

Fig. 3.2. Tray type gravity settling chamber

dust laden gas enters the cylindrical chamber centrifugal force is created by the action of circulator motion. The dust particle nearly 1000 time less denser than gas can not take up sudden change in direction on centrifugal impaction on the wall of the cylindrical chamber, the inertia of the particle is greatly reduced and trying fall on the conical chamber. The circulatory motion of the gas in the cylindrical section creates outer vortex and this vortex gradually reduced to in the conical section and the inner vortex is created at the end of the conical section. The gas in the inner vortex is free of dust and goes out in the outlet chamber. The efficiency of the cyclone is nearly about 70 -95% depending on the particle density and size. This device also offers low pressure drop in the range of 60-100 mm of water column.

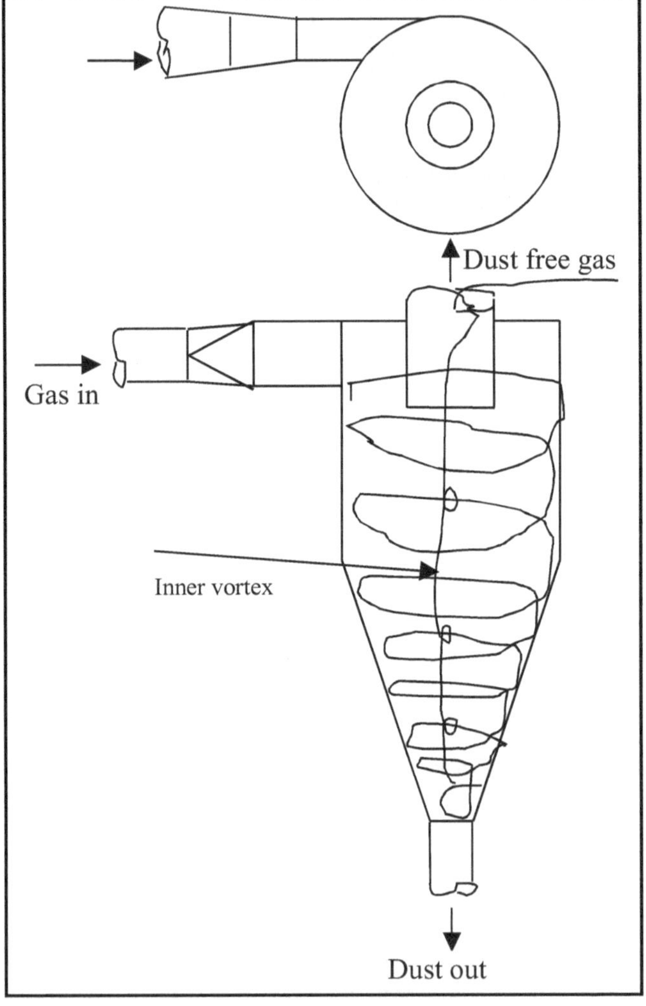

Fig. 3.3. Cyclone Separator

(iii) Fabric Filter

This is one of the oldest methods of removing dust particles in cement industries. In this device the dust laden gas passes through a fabric bag where particles are got intercepted and dust free gas passes out. The collected dust particle in the form of thin layer inside the bag will be periodically removed by mechanical ram of shaker. Fabrics are made of woven fabric or felt cloth made synthetic or cotton depending on the corrosive nature of the dust particles, operating temperature and pressure. The typical filter shown in Fig is tubular bag which is closed at the upper end and has a conical hopper attached the lower end to enable the dust particle to be collected. The bags varying in number arranged rows and column in rectangular chamber. Fabric filter have the advantages like low pressure drop, high collection efficiency up to 90%.

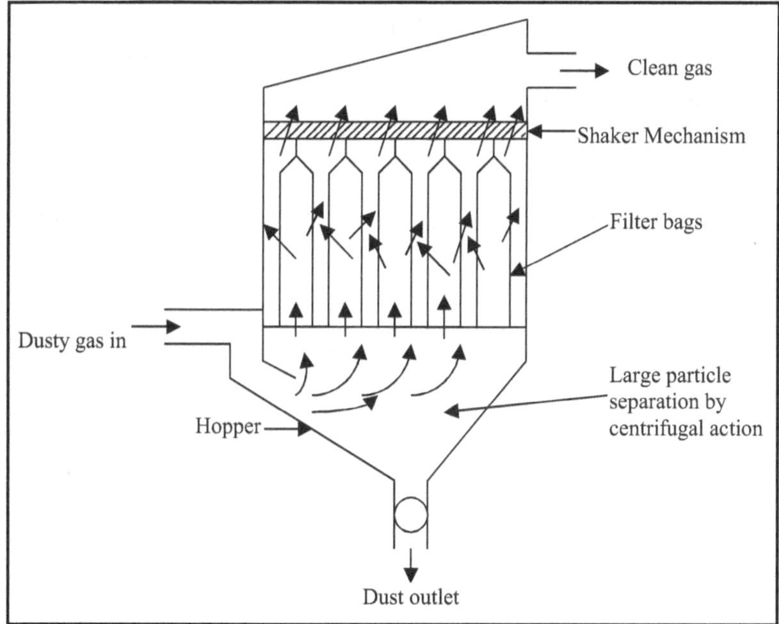

Fig. 3.4. Fabric or bag filter

(iv) Electrostatic Precipitator

Electrostatic precipitator is one of the very widely used dust collecting equipment with the efficiency up to 100%. The device is widely used in power plant, cement plant, paper mills and refineries. It is working on the principles of electrostatic charging and precipitation. This device consists of positively charged collecting surface plate which is grounded, and negatively charged series of electrode wire both of which are subjected to a D.C current at high voltage in the range of 50-60. Sometimes of a series of alternate collecting plate and wire s forming multi chamber is available to increase the capacity. These chambers are accommodated in the rectangular chamber at the top and conical chamber at the bottom just it was seen in the Fabric filter. The mechanical ram provided at the bottom periodically shake the collector plate by mechanical action and dust particle are getting collected. Fig shows typical single electrode electrostatic precipitator.

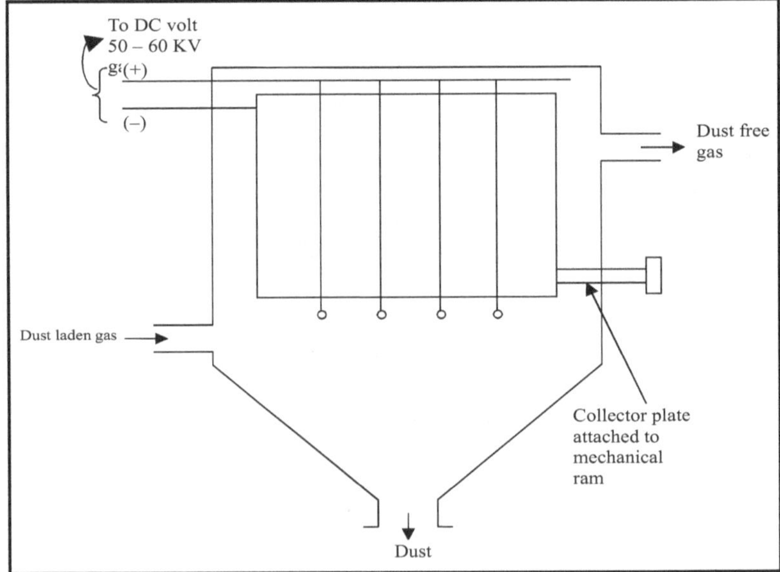

Fig. 3.5. Electrostatic precipitator

(v) Wet Scrubber

In this type of collectors particles are being subjected to fine droplets of liquid which aids in the separation of particles. There are three types like spray, venture, and cyclone called centrifugal separator.

Spray Type

In this device fine droplets of water is introduced with the help of fine spray nozzle as shown in Fig. As the dust laden gas flow upwards collided with fine droplets forming bigger size and settle under influence of gravity. The efficiency of collection depends on fineness of liquid droplet produce by the nozzle as well as dust particle density and size.

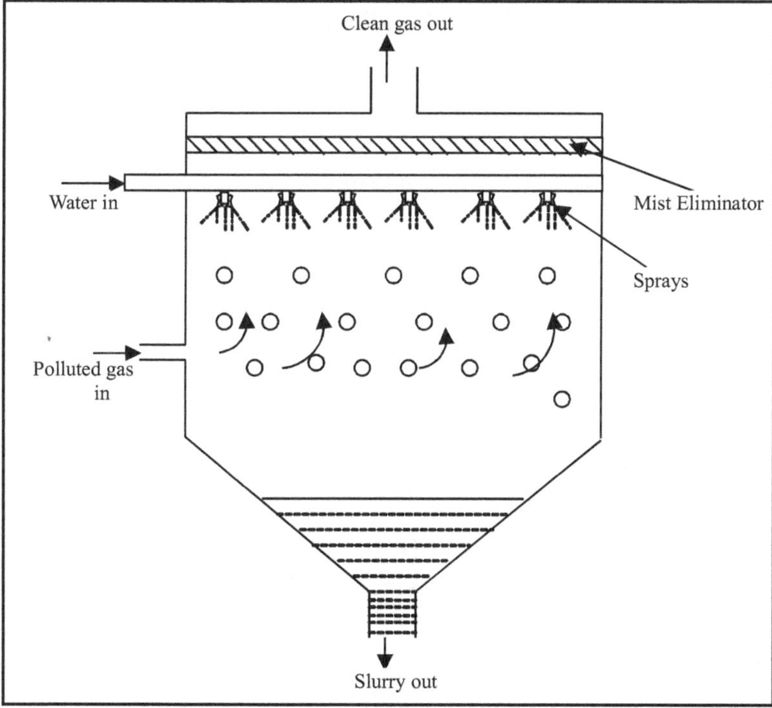

Fig. 3.6. Spray type tower

Centrifugal Separator

In this device dust laden gas enters the chamber which is provided with high pressure fine spray at the center. This fine droplet collided with dust particle in the outer vortex of gas in the chamber and on impact getting collected at the bottom.

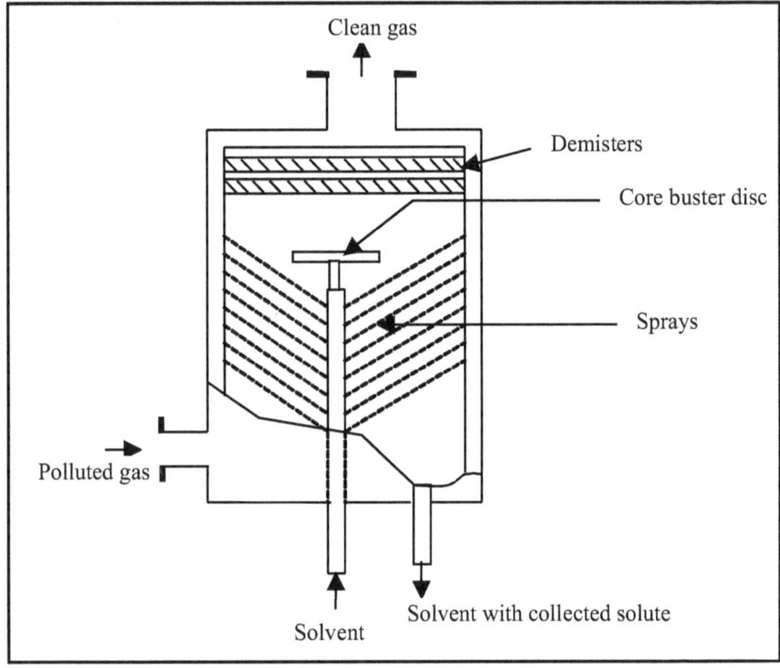

Fig. 3.7. Centrifugal spray type scrubber

Venturi Scrubber

This device offer high performance collection of fine particles of size less than 5 micron. The particle may be sticky, flammable or corrosive. The operating velocity of the scrubber is of the order of 50 to 150 m/s. It consists of venture type chamber with the throat section in between. Fine droplet of liquid being injected at the throats of the venture where there these droplets are mixed intimately with the high speed dust laden particle by agglomeration which is subsequently getting separated in the cyclone separator collector connected at the divergent section of the venture scrubber. Fig 3.8 show the schematic operation of this unit.

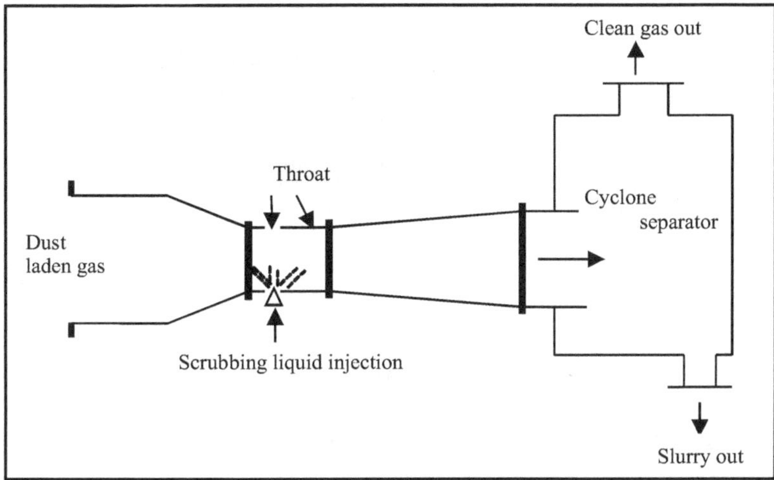

Fig. 3.8. Venturi Scrubber for dust removal

3.1.5 Gaseous Pollution Control Equipments

Removal of gaseous pollutants can be accomplished by two method namely absorption and adsorption. Absorption of can be done in simply by passing the gaseous pollutants in a highly soluble solvent so that gaseous pollutants will get absorbed or it can be passed through a packed / plate type of venture scrubber where high area of contact is being created for liquid and gaseous pollutants for efficient separation. A wide variety of solvents are available for the absorption of various gaseous pollutants. These are given in the Table 3.1. In adsorption, the pollutants are physically adsorbed on the solid and once the saturation limit adsorption is reached, the adsorbent will be regenerated by suitable gaseous stream.

Table 3.1. Solvents used for various gas pollutants

Name of the gaseous pollutants	Name of the solvent
CO_2	Carbonates
CO_2	Hydroxides
CO_2	Ethanolamine
CO	Cuprous amine complexes
CO	Cuprous Ammonium Chloride
SO_2	Calcium hydroxide
SO_2	Ozone-water
SO_2	$HCrO_4$
SO_2	KOH
Cl_2	Water
Cl_2	$FeCl_2$
H_2S	Ethanolamine
H_2S	$Fe(OH)_2$
SO_3	Sulfuric Acid
C_2H_4	KOH
C_2H_4	Trialkyl Phosphates
Olefins	Cuprous Ammonium Complexes
NO	Ferrous sulfate
NO	Calcium hydroxide
NO	Sulfuric Acid
NO_2	Water
NH_3	Water

(i) Packed Tower

This device consists of cylindrical chamber packed with packing made ceramic or plastic of various types depending corrosive action of the gaseous pollutants to be absorbed. The purpose of the packing is provide high area of contact between gas and solvent. The solvent is sprayed from the sprayer at the top over the packing. The pollutant laden gas is passed from the bottom through the packing which is wetted with solvent and getting absorbed and comes out.

The demister is provided above the sprayers to eliminate carry over fine droplets of solvent by the gas.

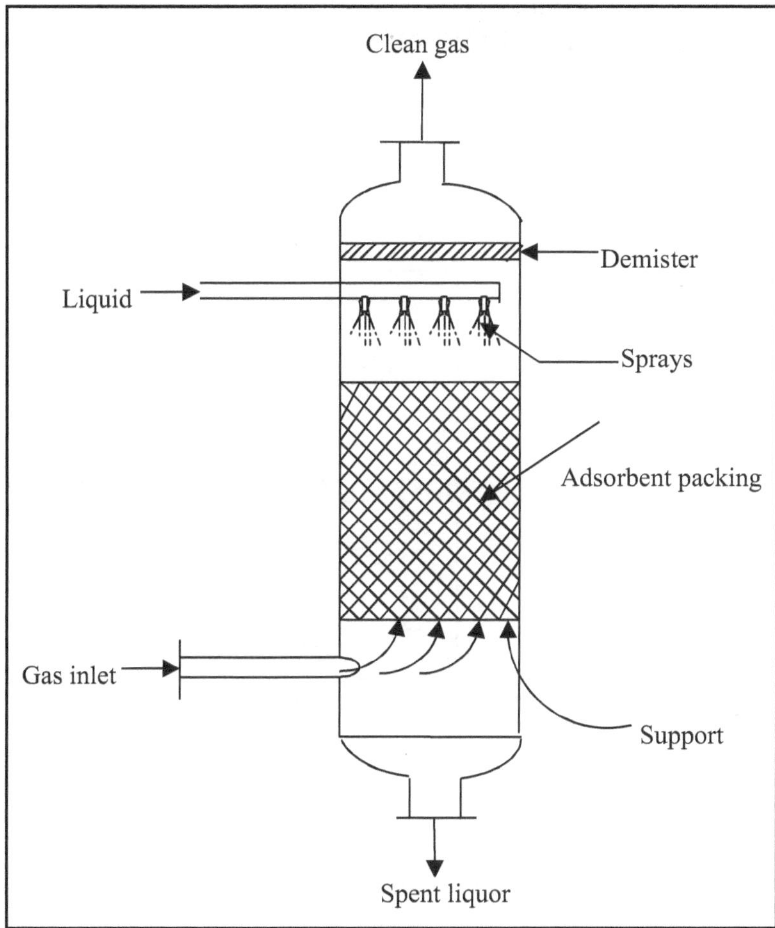

Fig. 3.9. Packed absorption tower for gaseous pollutants removal

(ii) Plate Tower

This device is less efficient that packed tower but offer low pressure drop. In this device there are a number sieve plate where there is a stagnant layer liquid through a pollutant gas is passed in a direction opposite to that of solvent. The industrial scale the will be 20 to 30 trays depending on the level of pollutants in the gas and degree removal of pollutants. The gas velocity will be such that to prevent carry over of solvent.

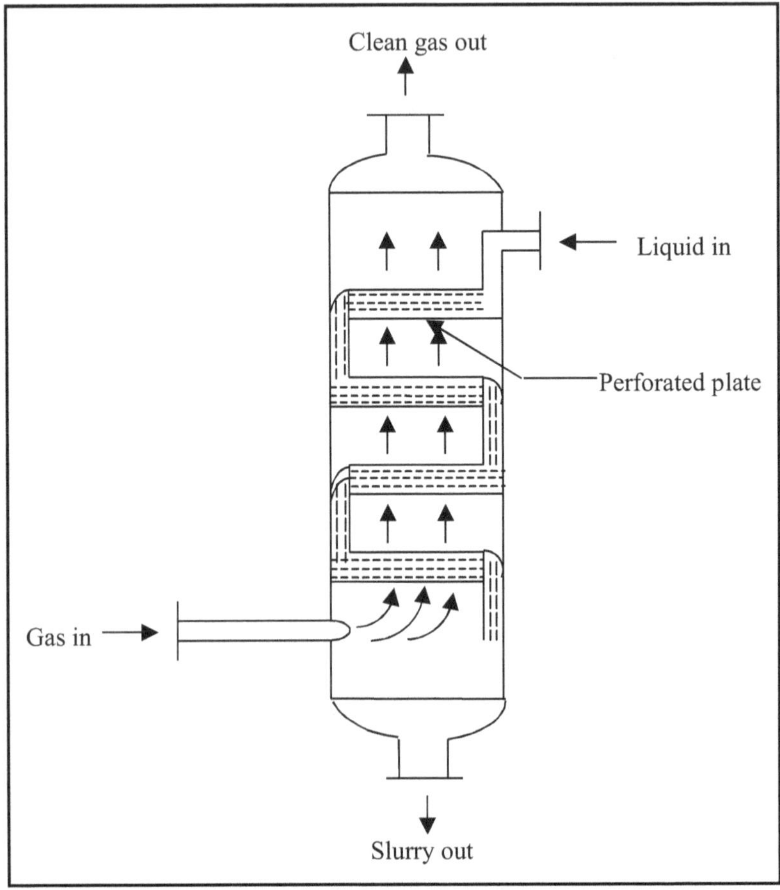

Fig. 3.10. Plate absorption tower for gaseous pollutants removal

(iii) Venturi Scrubber

This device is similar to venture scrubber used for dust particle removal except that the solvent is used in stead of water depending on the type of pollutant.

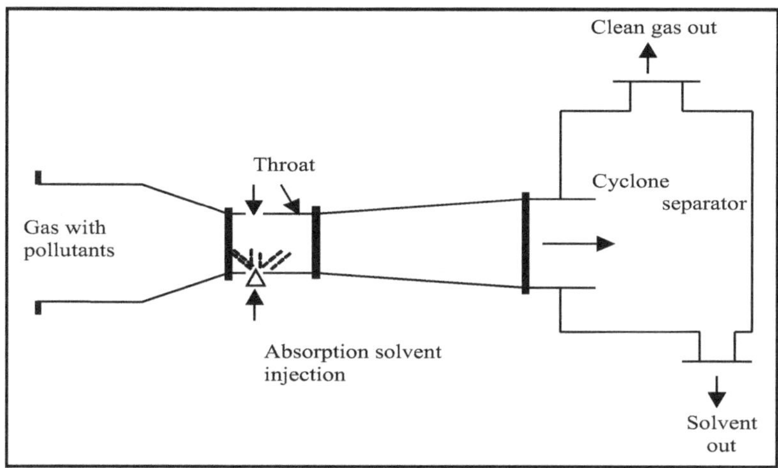

Fig. 3.11. Venturi type absorption unit

Adsorption

This device consists of cylindrical chamber packed with solid adsorbent. The adsorbent may activate, alumina, or silica based on the removal of pollutants. The pollutants laden gas is passed through packing where it is getting adsorbed. Once the maximum adsorption limit is reached the adsorbent is regenerated with suitable gaseous stream. Generally super heated steam is used for most organic gaseous pollutants. This steam is passed through the adsorbent and adsorbed pollutant will be removed from adsorbent and another cycle of adsorption will be done. This kind operation will be cyclically done in a batch manner and will be so efficient as absorption. It will be most suitable for small scale and low level pollutants present in the gas stream.

3.2 Water Pollution

Water pollution occurs when a body of water is adversely affected due to the addition of large amounts of materials to the water. When it is unfit for its intended

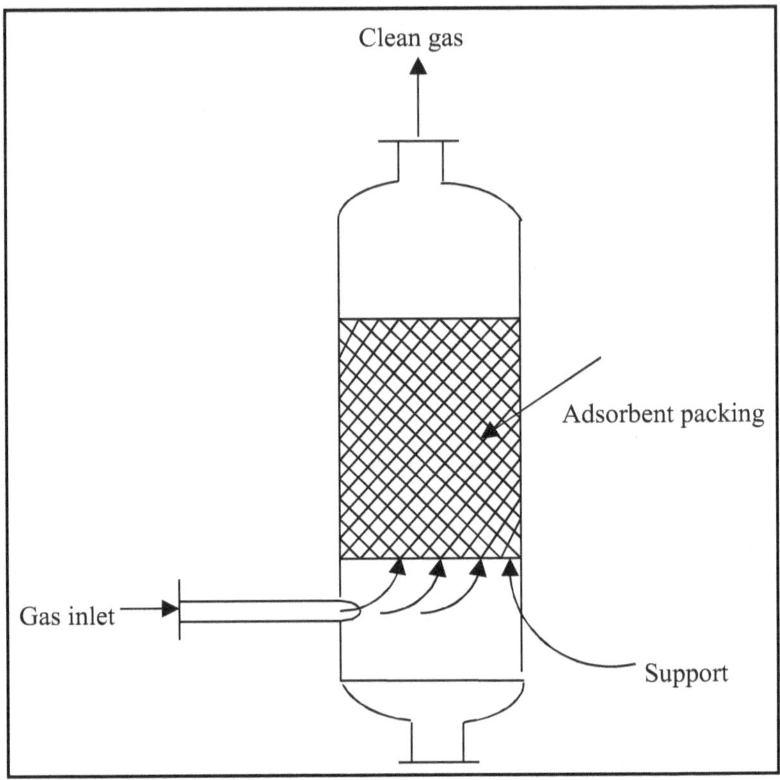

Fig. 3.12. Adsorption tower for gaseous pollutants removal

use, water is considered polluted. Two types of water pollutants exist; point source and non-point source. Point sources of pollution occur when harmful substances are emitted directly into a body of water. A non-point source delivers pollutants indirectly through environmental changes. An example of this type of water pollution is when fertilizer from a field is carried into a stream by rain, in the form of run-off which in turn effects aquatic life. The technology exists for point sources of pollution to be monitored and regulated, although political factors may complicate matters. Non-point sources are much more difficult to control. Pollution arising from nonpoint sources accounts for a majority of the contaminants in streams and lakes.

There are sources of water pollution can be classified as municipal, industrial, and agricultural. Municipal water pollution consists of waste water from homes and commercial establishments. For many years, the main goal of treating municipal wastewater was simply to reduce its content of suspended solids, oxygen-demanding materials, dissolved inorganic compounds, and harmful bacteria. In recent years, however, more stress has been placed on improving means of disposal of the solid residues from the municipal treatment processes. The basic methods of treating municipal wastewater fall into three stages: primary treatment, including grit removal, screening, grinding, and sedimentation; secondary treatment, which entails oxidation of dissolved organic matter by means of using biologically active sludge, which is then filtered off; and tertiary treatment, in which advanced biological methods of nitrogen removal and chemical and physical methods such as granular filtration and activated carbon absorption are employed. The handling and disposal of solid residues can account for 25 to 50 percent of the capital and operational costs of a treatment plant. The characteristics of industrial waste waters can differ considerably both within and among industries. The impact of industrial discharges depends not only on their collective characteristics, such as biochemical oxygen demand and the amount of suspended solids, but also on their content of specific inorganic and organic substances. Three options are available in controlling industrial wastewater. Control can take place at the point of generation in the plant; wastewater can be pretreated for discharge to municipal treatment sources; or wastewater can be treated completely at the plant and either reused or discharged directly into receiving.

3.2.1 Causes of Pollution

Many causes of pollution including sewage and fertilizers contain nutrients such as nitrates and phosphates. In excess levels, nutrients over stimulate the growth of aquatic plants and algae. Excessive growth of these types of organisms consequently clogs our waterways, use up dissolved oxygen as they decompose, and block light to deeper waters. This, in turn, proves very harmful to aquatic organisms as it affects the respiration ability or fish and other invertebrates that reside in water.

Pollution is also caused when silt and other suspended solids, such as soil, wash off plowed fields, construction and logging sites, urban areas, and eroded river banks when it rains. Under natural conditions, lakes, rivers, and other water bodies undergo Eutrophication, an aging process that slowly fills in the water body with sediment and organic matter. When these sediments enter various bodies of water, fish respiration becomes impaired, plant productivity and water depth become reduced, and aquatic organisms and their environments become suffocated. Pollution in the form of organic material enters waterways in many different forms as sewage, as leaves and grass clippings, or as runoff from livestock feedlots and pastures. When natural bacteria and protozoan in the water break down this organic material, they begin to use up the oxygen dissolved in the water. Many types of fish and bottom-dwelling animals cannot survive when levels of dissolved oxygen drop below two to five parts per million. When this occurs, it kills aquatic organisms in large numbers which leads to disruptions in the food chain.

Pathogens are another type of pollution that prove very harmful. They can cause many illnesses that range from typhoid and dysentery to minor respiratory and skin diseases. Pathogens include such organisms as bacteria, viruses, and protozoan. These pollutants enter waterways through untreated sewage, storm drains, septic tanks, runoff from farms, and particularly boats that dump sewage. Though microscopic, these pollutants have tremendous effect evidenced by their ability to cause sickness.

3.2.2 Origin of Wastewater

Wastewaters can be classified by their origin as domestic wastewater and industrial wastewater. Any combination of wastewaters that is collected in municipal sewers is termed as municipal sewage. Domestic wastewater is that which is discharged from residential and commercial establishments, whereas industrial wastewater is that which is discharged from manufacturing plants. The pollutants in domestic wastewater arise from residential and commercial cleaning operations, laundry, food preparation, body cleaning functions and body excretions. The composition of domestic wastewater is relatively constant.

Industrial wastewater is formed at industrial plants where water is used for various processes and also for washing and rinsing equipment, rooms, and etc. These operations result in the pollution of nearby aquatic systems because some of the products and byproducts are discharged, either deliberately or unintentionally into them.

Normally, wastewaters are conducted to treatment plants for removing undesirable components, which include both organic and inorganic matter as well as soluble and insoluble material. These pollutants, if discharged directly or with improper treatment, can interfere with the self-cleaning mechanisms of water bodies. The capacity for self - cleaning is due to the presence of relatively small numbers of different types of micro - organisms in the water bodies. These micro - organisms use as food much of the organic pollutants and break them down into simple compounds such as CO_2 or methane, and the micro - organisms produce new cells also. But often either a pollutant does not degrade naturally or the sheer volume of the pollutant discharged is sufficient to overwhelm the self - cleaning processes. Also, the microbial population can be destroyed by toxic wastes discharged into the waterway. If that happens, the pollutant concentrations will build up and reach high enough levels that will prevent re - establishment of a microbial population. The water quality thus becomes permanently degraded.

Various constituents of wastewater are potentially harmful to the environment and to human health. In the environment, the pollutants may cause destruction of animal and plant life and aesthetic nuisance. Drinking water sources are often threatened by increasing concentration of pathogenic organisms as well as by many of the new toxic chemicals disposed of by industry and agriculture. Thus, the treatment of these wastes is of paramount importance.

3.2.3 Waste Water Composition

Depending on the amounts of physical, chemical and biological constituents of wastewater, they may be classified as strong, medium, or weak.

Typical compositions for domestic wastewater are given in Table No 2. The compositions and concentrations are highly variable and hence, the table is intended only to serve as a guide and not as a basis for design data on typical characteristics of municipal sewage in some urban cities of India are given in the following:

- All values except for settleasble solids are expressed in mg/l.
- Values should be increased by amount in carriage

The composition of industrial wastewater is quite varied, its constituents ranging from organic solvents, oils, suspended solids to dissolved chemical compounds. Table lists a number of potentially polluting chemical substances released by different industries. Most of the chemicals are suspect agents" include 4 - nitrodiphenyl, 4 - aminodiphenyl, - and – napthalamine, methyl chloromethyl ether, benzidine 3, 3′ - dichlorobenzidine, - propiolactone, 4 – dimethylaminoazobenzene, N – nitro sodium ethylamine and vinyl chloride. Another important aspect of industrial wastewaters is that some of the chemicals are valuable enough to warrant recovery. This aspect of recovery has taken on a new perspective in recent years and the recovered substance can either be recycled within the plant, or offered to other companies for use as basic of intermediate chemicals.

Table 3.2. Typical composition of domestic wastewater

Constituent	Strong	Concen-tration * Medium	Weak
Solids, total	1200	700	350
Dissolved solid, total	850	500	250
Fixed	525	300	145
Volatile	325	200	105
Suspended solids, total	350	200	100
Fixed	75	50	30
Volatile	275	150	70
Settle able solids (ml/l)	20	10	5
Biological oxygen demand 5 days, 20° C	300	200	100
Total organic carbon (TOC)	200	135	665
Chemical oxygen demand (COD)	1000	500	250
Nitrogen (total as N)	85	40	20

Contd...

Table 3.2. Contd...

Organic	35	15	8
Free ammonia	50	25	12
Nitrites	0	0	0
Nitrates	0	0	0
Phosphorous (total as P)	20	10	6
Organic	5	3	2
Inorganic	15	7	4
Chlorides**	100	50	30
Alkalinity (as $CaCO_3$) * *	200	100	50
Cheese	150	100	50

* All values except for settleable solids are expressed in mg/l.

* Values should be increased by amount in carriage water.

3.2.4 Types of Water Polutants

The problem of water pollution due to discharge of domestic and industrial wastes into aquatic systems has already become a serious problem in the country. Nearly 75 to 80% of India's population is exposed to unsafe drinking water. As a result, enteric diseases, often reaching epidemic proportions, devastate several parts of the country. The rivers and lakes near urban centers emit disgusting odors and fish are being killed in millions along seacoasts. The meat of some of them is tainted and unsafe to eat because of excessive levels of mercury and pesticides in their bodies. The origin of these problems must be attributed to many sources and types of pollutants. Some pollutants may have indirect effects whilst substances normally not considered, as pollutants may become so under special circumstances. To aid in a systematic discussion of water pollutants, they have been classified into nine categories as described below:

(i) Oxygen demanding wastes
(ii) Disease - causing agents

(iii) Synthetic organic compounds
(iv) Plant nutrients
(v) Inorganic chemicals and minerals
(vi) Sediments
(vii) Radioactive substances
(viii) Thermal discharges
(ix) Oil

Pollution of the waterways is often caused by a combination of the above categories, which can severely compound the problem. Because of their vital importance in water quality control programmes, these categories are discussed briefly in the following sections.

3.2.5 Basic Processes of Water Treatment

The wastewater treatment processes are generally grouped according to the water quality they are expected to produce. These processes are usually grouped as the primary treatment, the secondary treatment, and the tertiary or the advanced waste treatment. Primary treatment removes identifiable suspended solids and floating matter. In the secondary treatment, also known as the biological treatment, organic matter that is soluble or in the colloidal form is removed. Advanced waste treatment may involve physical, chemical or biological processes or their various combinations depending on the impurities to be removed. These processes are employed to remove residual soluble non-biodegradable organic compounds, including surfactants, inorganic nutrients and salts, trace contaminants of various types, and dissolved inorganic salts. The advanced waste treatment processes are expensive and are used only when water produced is required to be of high quality than that produced by conventional secondary treatment so that the treated water can be reclaimed and put to some form of direct reuse.

In industrial scale after the primary treatment, the major chemical components present in the waste water is treated by adding suitable chemicals so that the waste can safely treated for subsequent biological treatment without shock loading. Usually biological treatment of any type generally used for removing small concentration of pollutants.

(i) Physical (Primary) Treatment

Primary treatment is to reduce oils, grease, fats, sand, grit, and coarse (settleable) solids whicl will affect subsequent treatment process. In some situations this treatment is also called as physical treatment as only physical operation are involved in this process.

Removal of Large Objects

In this process waste water is strained to remove all large objects that are deposited in the sewer system, such as rags, sticks, and other solid objects, cans, fruit, etc. This is most commonly done using a manual or automated mechanically raked screen. This type of waste is removed because it can damage the sensitive equipment in the waste water treatment plant.

Sand and Grit Removal

Sand grit and stones have to be removed early in the process to avoid damage to pumps and other equipment in the remaining treatment stages. Sometimes there is a sand washer (grit classifier) followed by a conveyor that transports the sand to a container for disposal. The contents from the sand catcher may be fed into the incinerator in a sludge processing plant but in many cases the sand and grit is sent to a land-fill.

Screening

The waste water after grid removal is then sent through fixed or rotating screens to remove floating and larger material such as rags and smaller particulates such as peas and corn. Screenings are collected and may be returned to the sludge treatment plant or may be disposed of off site by landfilling or incineration. Sometimes solids are cut into small particles through the use of rotating knife edges mounted on a revolving cylinder, is used in plants that are able to process this particulate waste. Figures shown below show the Manually and Mechanically raked screen

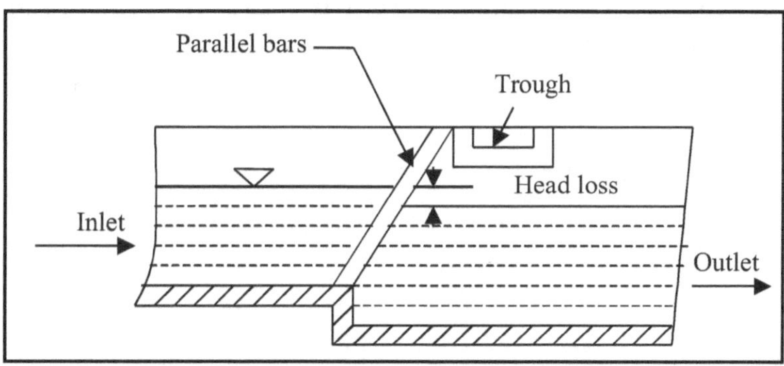

Fig. 3.13. Manually– raked bar screen

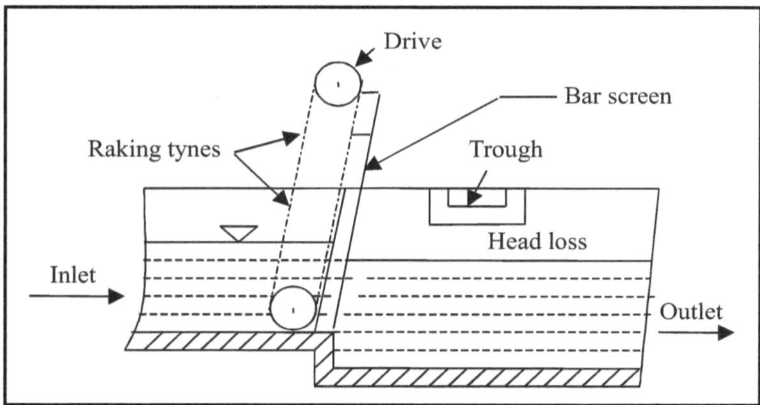

Fig. 3.14. Mechanically – raked bar screen

Sedimentation

The waster after screening is sent to the sedimentation tank which are large circular or rectangular or square tanks. The tanks are large enough that fine solids can settle and floating material such as grease and plastics can rise to the surface and be skimmed off. Primary settlement tanks are usually equipped with mechanically driven scrapers that continually drive the collected sludge towards a hopper in the base of the tank from where it can be pumped to further sludge treatment stages. Sedimentation tank usually are rectangular horizontal flow, Circular radial flow and Vertical up flow basin. The first two are used for higher capacity treatment and third one is being used for lower capacity.

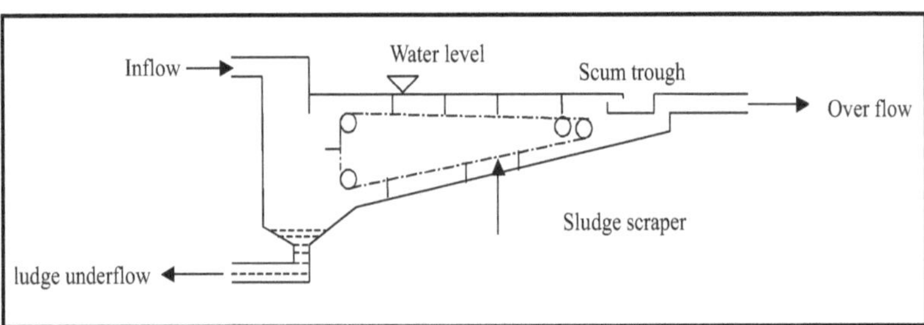

(a) Rectangular Horizontal flow type

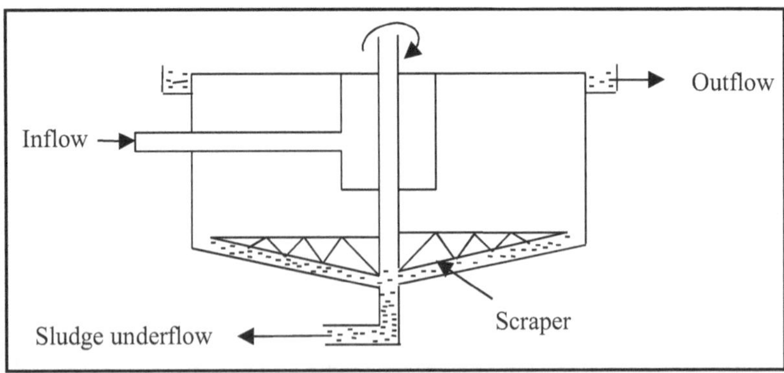

(b) Circular radial flow type

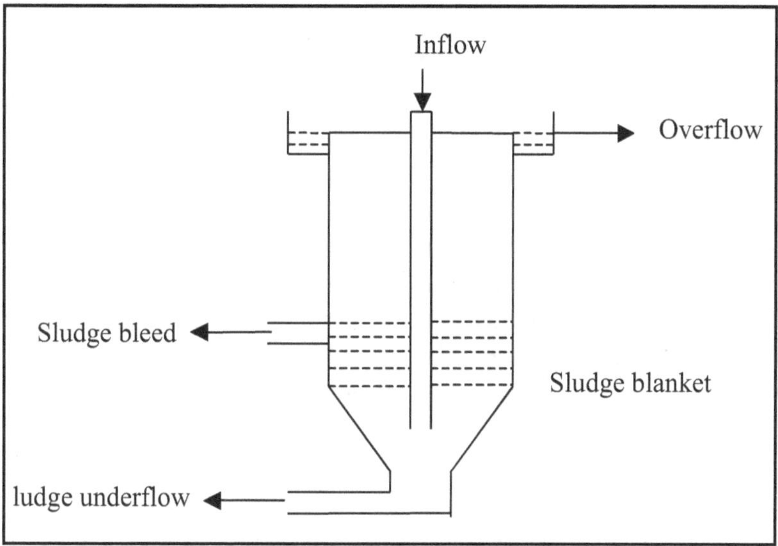

(c) Circular vertical up flow type

Fig. 3.15. Settling tank

Separation of Oil and Grease

Wastewaters contain variable amounts of oil and grease which depend on the process used, the species processed, and the operational procedure. To remove oil and grease, gravity separation may be used, provided the oil particles are large enough to float towards the surface and are not emulsified.

Flotation is an operation that removes not only oil and grease but also suspended solids. It is discussed in this section since it is one of the most effective systems for suspensions which contain oil and grease. The most common procedure is that of dissolved air flotation (DAF), in which the waste stream is first pressurized with air in a closed tank. After passing through a pressure-reduction valve, the wastewater enters the flotation tank where, due to the sudden reduction in pressure, minute air bubbles in the order of 50- 100 microns in diameter are formed. As the bubbles rise to the surface, the suspended solids and oil or grease particles adhere to them and are carried upwards. It is common practice to use chemicals to enhance flotation performance.

(ii) Biological (Secondary) Treatment

The water leaving the primary process has lost much of the solid matter but still contains a high demand for oxygen i.e., it is composed of high-energy molecules, which will be decomposed by microbial action, thus creating a biochemical oxygen demand (B.O.D.). This demand for oxygen must be reduced (energy wasted) if the discharge is not to create unacceptable conditions in the watercourse. The objective of secondary treatment is thus to remove BOD.

All secondary methods by definition use microbial action to reduce the energy level (BOD) of the waste. Although there are many ways the microorganisms can be put to work, the first really successful modern method of secondary treatment was the "trickling filter".

(a) Trickling Filter

A trickling filter is a bed of solid media for bacteria to attach on its surfaces. Wastewater is irrigated on the solid media (Figure). It is also called a biological filter to emphasize that the filtration process is not mechanical straining of solids, but removal of organic substances by use of bacterial action.

The solid media can be stones, waste coal, gravel or specially manufactured plastic media. The latter can be corrugated plastic sheets or hollow plastic cylinders, with the main aim being to provide a large surface area for bacteria to attach to, while at the same time allowing free movement of air. Typically the solid media is placed in a

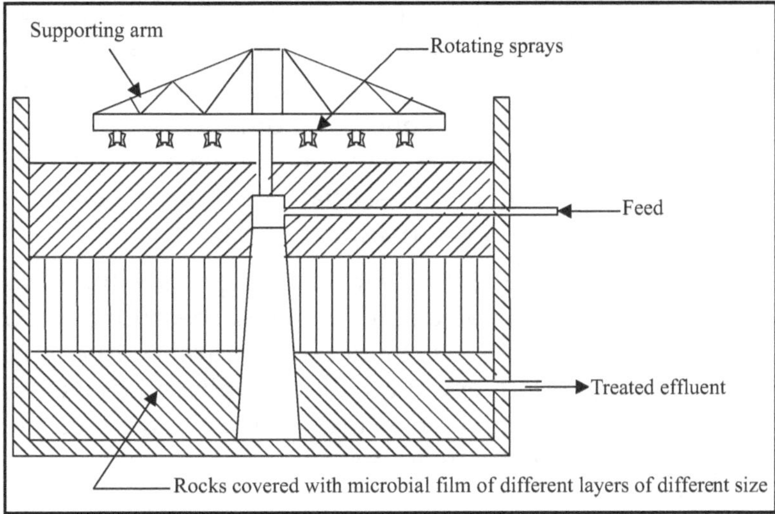

Fig. 3.16. Trickling Filter for biological treatment of astewater

tank on a support with openings to allow air to move up by natural convection and for treated wastewater to be collected in the under-drain.

Wastewater has to undergo primary treatment before trickling filtration; otherwise solids will block the filter. As wastewater trickles over the surfaces of the solid media organic substances are trapped in the layer of bacterial slime. The bacteria consume the organic substances in the same manner as in the activated sludge process, while air diffuses into the slime layer from the air spaces in the bed of the trickling filter. Growth and reproduction of the bacteria take place and result in an increase of thickness of the slime layer, particularly at the top of the biological filter. Periodically bacterial slime sloughs off the surfaces of the filter media and leaves with the treated wastewater.

Solids derived from the sloughing off of bacterial slime are separated from the treated wastewater in a sedimentation tank. Sludge from this sedimentation tank is not returned to the trickling filter, but treated prior to reuse or disposal . Treated wastewater can however be returned to the trickling filter, if this will assist with either treating the wastewater further (second pass) or more generally for a more uniform distribution of water over the trickling filter bed. The trickling filter and associated sedimentation tank is also termed 'secondary treatment'.

The energy requirement for operating a trickling filter is less than for an activated sludge process, because oxygen supply to the bacteria is provided by natural diffusion of air. The area requirement of a biological filter is, however, larger than for an activated sludge process to achieve the same quality of treated wastewater.

(b) Activated Sludge Treatment

The term 'activated sludge' refers to sludge in the aeration tank of an activated sludge treatment process. It consists of flocs of bacteria, which consume the biodegradable organic substances in the wastewater. Because of its usefulness in removing organic substances from wastewater, the sludge is kept in the process by separating it from the treated wastewater and re-circulating it. A typical arrangement of an activated sludge process is schematically shown in Figure.3.17

The primary-treated wastewater is then passed to an aeration chamber. Aeration provides oxygen to the activated sludge and at the same time thoroughly mixes the sludge and the wastewater. Aeration is by either bubbling air through diffusers at the bottom of the aeration tank, or by mechanically agitating the surface of the water.

In the aeration tank, the bacteria in the activated sludge consume the organic substances in the wastewater. The organic substances are utilized by the bacteria for energy, growth and reproduction. The wastewater spends a few hours in the aeration chamber before entering a second sedimentation tank to separate the activated sludge from the treated wastewater. The activated sludge is returned to the aeration tank. There is an increase in the amount of activated sludge because of growth and reproduction of the bacteria. The excess sludge is wasted to maintain a desired amount of sludge in the system. This part of the treatment process is called 'secondary treatment', the sedimentation tank as secondary sedimentation tank, the overflow from the sedimentation tank as secondary-treated wastewater (secondary effluent) and the excess activated sludge as secondary sludge.

Depending on the flow rate of wastewater, several parallel trains of primary and secondary stages can be employed. There are several ways to operate an activated sludge process. In a 'high rate' process a relatively high volume of wastewater is treated per unit volume of activated sludge. The high amount of organic waste consumed by the activated sludge produces a high amount of excess sludge. In an 'extended aeration' mode of operation the opposite condition takes place. A relatively low amount of organic waste is treated per unit volume of sludge with little excess sludge to be removed. Removal of BOD is higher in the extended aeration mode compared to the high rate mode, but more wastewater can be treated with the latter mode.

An activated sludge treatment plant is a highly mechanized plant, and is suited to automated operation. The capital cost for building such a plant is relatively high. The energy requirement, particularly for providing air to the aeration tank, is also relatively high. There is a need for regular maintenance of the mechanical equipment, which requires skilled technical personnel and suitable spare parts. The operation and maintenance costs of an activated sludge treatment plant are therefore relatively high.

An activated sludge treatment process can be operated in batches rather than continuously. One tank is allowed to fill with wastewater. It is then aerated to satisfy the oxygen demand of the wastewater, following which the activated sludge is allowed to settle. The treated wastewater is then decanted, and the tank is filled with a new batch of wastewater. At least two tanks are needed for the batch mode of operation, constituting what is called a 'sequential batch reactor (SBR)'. SBRs are suited to smaller flows, because the size of each tank is determined by the volume of wastewater produced during the treatment period in the other tank.

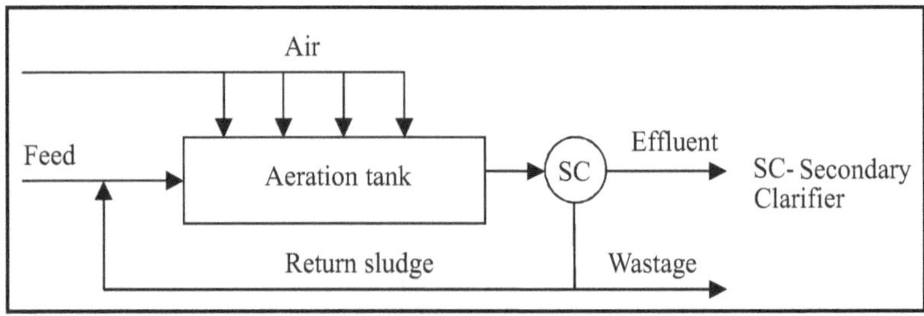

Fig. 3.17. Scheme of Activated Sludge Treatment Process

(c) Rotating Biological Contractors

Rotating biological contractors (RBC) units are another form of attached growth processes. In RBC units the biomass is attached to disks with the diameter up to 3.5 m. The drum rotate at 1 to 3 rpm while immersed up to 40% in the wastewater The disks are made of corrugated, light plastic material or light smooth steel material. When exposed to air the attached biomass absorbs air and when immersed the microorganisms absorb the organic load. A biomass of 1-4 mm grows on the surface and its excess is teared off the disks by shearing forces and is separated from the liquid in the secondary settling tank. A small portion of the biomass remains suspended in the liquid within the basin and is also responsible in minor part for the organic load removal. Optimum speed of rotation is 2 to 3 rpm Rotation speeds of more than 3 rpm are seldom used because this increases electric power consumption while the oxygen transfer does not increase sufficiently. The ratio of surface area of disks to liquid volume is typically 5 l/m^2. For high-strength wastewaters, more than one unit in series also called staging is used. The effect of lower temperatures is partially mitigated by the use of housing for the disk units. These systems are normally operated without recycling the liquid. The power consumption is in the order of 2 kW/1 000 m^3/ day of capacity.

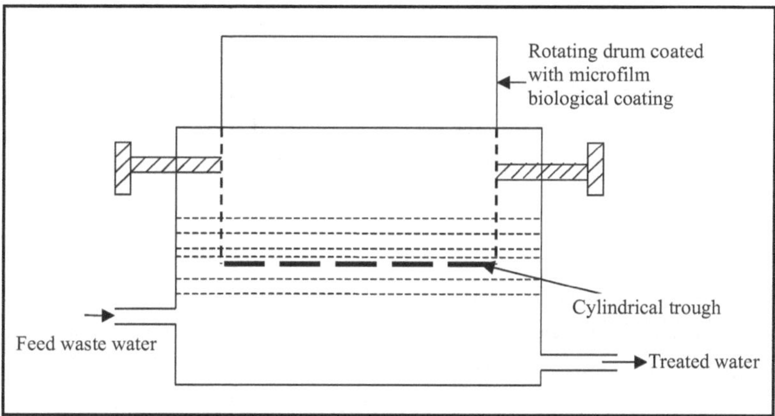

Fig. 3.18. Rotating biological disc contactor

Selection of Aerobic Treatments

Several factors influence the choice of a particular aerobic treatment system besides economics. There is no universal solution and the decision of which system to use (or even if using an aerated system or not) depends on many aspects. Main factors are the area available, which sometimes is the deciding aspect; the ability to operate intermittently is critical for several fishing industries which do not operate in a continuous fashion or work only seasonally; the skill needed for operation of a particular treatment cannot be neglected; and finally the costs (both operating and initial investment) are also sometimes decisive.

(iii) Tertiary Treatment

The effluent from atypical secondary treatment plant still contains 20-40 mg/l suspended solids and 20-40 mg/l BOD, which may be objectionable in some streams. Suspended solids, in addition to contributing to BOD, may settle on the streambed and inhibit certain forms of aquatic life. The BOD, if discharged into a stream with low flow, can cause damage to aquatic life by reducing the dissolved oxygen content. In addition, the secondary effluent contains significant amounts of plant nutrients and dissolved solids. If wastewater is of industrial origin, it may also contain traces of organic chemicals, heavy metals and other contaminants.

The recent trend towards the formulation of regulation for the discharge of specific compounds and the increased emphasis on recovery of valuables from industrial wastewater has created need for treatment beyond the conventional secondary treatment stage. Tertiary treatment processes are expensive at the present level of their development. Their need in a particular situation should, therefore, be assessed in the light of circumstances relevant to that situation.

A wide variety of methods are used in advanced waste treatment to satisfy any of several goals, which include the removal of (1) suspended solids, (2) BOD, (3) plant nutrients, (4) dissolved solids and (5) toxic substances.

These methods may be introduced at any stage of the total treatment process as in the case of industrial wastewaters or may be used for complete removal of pollutants after the secondary treatment. In this section only a few representative methods are discussed to illustrate the problems and the potential solutions associated with tertiary wastewater treatment.

(iv) Types of Tertiary Treatment

Several methods now exist for the tertiary treatment of sewage notably:

(a) Rapid gravity sand filters
(b) Micro strainers
(c) Slow sand filters
(d) Irrigation over grass plots
(e) Lagoons
(f) Upward flow clarifiers
(g) Activated carbon methods

(a) Rapid Sand Filters

Rapid gravity sand filters are similar to those used in water treatments. These consist of a layer of graded sand 0.9 to 1.7 mm size and 0.6 m deep supported on a layer of gravel resting on a filter floor fitted with nozzles. The rate of filtration varies from 5000 to 8000 litters per sq.m.per hour. Such filters have shown a suspended solids reduction of 72-91 % and a BOD reduction of 52-70%.

It is desirable to backwash the filters at least once a day and use also air-scour in the process so that the accumulations of organic growth in the medium can be avoided. Alternatively, filter beds may also be required to heavily chlorinated or purged, regularly with caustic or other chemicals.

Once the backwash will stratify the sand layer such that only the top layer of the sand will be effective in filtration, upward flow sand filters have been used in which the whole depth of the filter medium will be effective. These are called "immedium" filters where higher rates of flow up to 16000 liters per sq.m.per. hour are possible.

(b) Micro Strainers

Micro strainers originally employed for raw water supply, have been used for tertiary treatment of sewage. It may consist of a drum revolving about a horizontal axis into one end of which influent is admitted. The sewage passes through strainer openings of size 25 to 150 microns made up of special stainless steel fabric on the cylindrical surface of the drum. A row of jets mounted on the top of the drum delivers the effluent under pressure for removing the solids from the drum as it rotates. A high intensity ultraviolet lamp is also mounted alongside the wash water jets in order to prevent the formation of biological growth on the fabric.

Suspended solids removal of 68-77 % and BOD removal up to 52 % have been reported for treated effluents. However, it may be mentioned that this micro strainer cannot be effectively used for relatively untreated sewage, since it gives rise to the problems of blocking of the openings due to biological growth, which the ultra violet lamp cannot prevent.

Filters using semi-permeable membranes having pores in the range of 0.025 to 5.0 microns called "ultra-filtration" membranes are being tried for the tertiary treatment of sewage when 70 to 90 % removal of organic matters in settled sewage has been reported. However, this process required further development for economical and wide operation.

(c) Slow Sand Filters

The construction and operation of slow sand filters are much simpler than those of rapid sand filters. The sewage passes through the filter at a much lower rate and when clogging occurs, the filter id drained and partially dried and the top layer of sand removed before filtration is continued. Results have shown a suspended solids removal of 60 % and a BOD removal of 40 %.

Under favorable geological and topological conditions natural sand deposits can be converted to intermittent sand filter operations as described above. The rates of loading on slow sand filters are of the order of 100 liters per sq.m.per. hour.

(d) Grass Plots

Irrigation of effluent over grassland has been used successfully at many works for final treatment. In this method the effluent is run to land through a system of channels and is collected by a second series of channels. Grass plots of slope not greater than 1 in 60 may be used to grow deep-rooted grasses. The treatment show a suspended solids removal of about 68 % and a BOD removal of 56 % with hydraulic loadings up to 1.34 million liters per hectare per day. Occasional cutting of the grass and periodical cleaning of the plots for the removal of accumulated solids have to be carried out.

(e) Lagoons

Method

Storage lagoon is another method, which is being used very successfully for improving effluents, the purification occurring by a combination of sedimentation and biological action.

Construction and Operation

It has been reported that when sewage effluent of about 7 ml passed through a series of three lagoons varying in depth from 2.0 to 4.0 m constructed in gravelly soil with an average retention time of 17 days, it produced a suspended solid reduction of average 83 % while the BOD reduction varied between 62 to 94 %

There was also a significant reduction in the members of E-Coli by about 99.5 %. An outstanding feature of the lagoon treatment was the production of a fully oxygenated effluent to discharge into receiving bodies of water so as to avoid any harmful effect on the same over if the dilution below.

Single lagoons with shallow depths and short retention periods were found to give satisfactory results whereas lagoons in series with larger depths and longer retention periods were highly effective.

An excess of algae, which may be passed from such lagoons, may be controlled by growing a variety of fish in the lagoons. Lagoons may also provide an open area as a habitat for a wide variety of wild life mainly birds. Therefore, lagooning of nitrified effluent is a tertiary treatment process, which has proved to be technically reliable and economically sound, and which adds to the quality of life.

Investigations carried out in India indicate that the algae that are responsible for the biological stabilization in lagoons are a succession of algae in sewage which vary from green algae to blue green algae with a corresponding increase in nitrate concentration in sewage, indicating thereby a favorable tertiary treatment.

(f) Upward Flow Clarifier

Method

The method consists of the clarification of the final effluent by upward flow through a shallow bed of gravel.

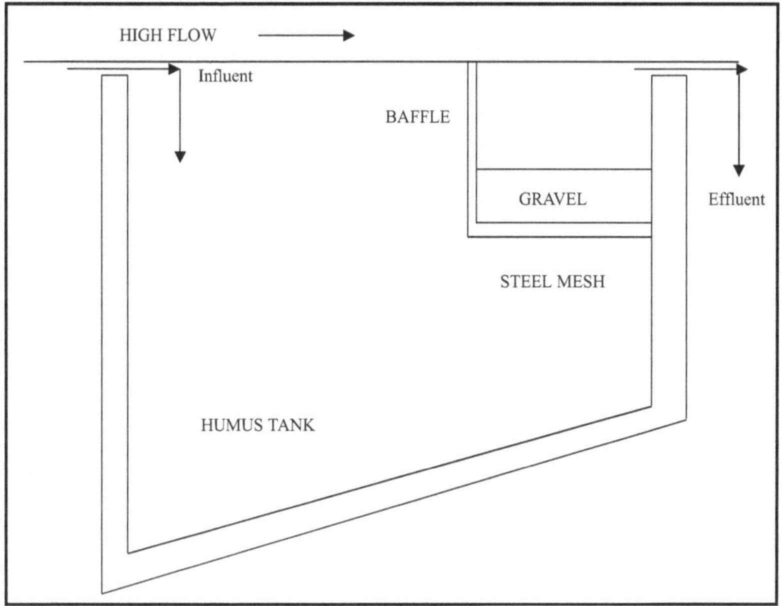

Fig. 3.19. Upward flow Clarifier

The humus tank effluent is passed through a 15 cm layer of gravel or pebbles of size 6 mm to 9 mm supported on a perforated floor either inside the humus tank or in a special compartment, the flow being in an upward direction at a rate of 600 to 1000 liters per hour. Refer to FIG.

The solids accumulated in the bed are removed from time to time by backwashing i.e., by lowering the liquid level by running off the effluent from below the pebble-bed and this process may be completed by washing the drained bed with a jet of water or effluent.

It has been determined by tests that depth and size of the bed and rate of flow as given above are the optimum values. The minimum depth of the liquid above the bed was also determined to about 0.3 m for effective performance.

The upward flow clarifier has a suspended solids removal efficiency of 60 % and a BOD removal of up to 40%.

(g) Activated Carbon Method

Method

The property of activated carbon to remove organic impurities by absorption is employed in water treatment. This is essentially a non-biological process and can be used for sewage also.

The very high specific surface of the material and the active properties of the interface are responsible for the absorptive properties. The absorptive capacity increases with porosity and consequently its ability to with stand repeated generation processes decreases.

Since the ability of activated carbon is reduced under alkaline conditions, it will be necessary to neutralize alkaline settled sewage to $_pH$ 7.0 before the adsorption process.

The passage of partially treated waste water containing suspended material through a bed of granulated activated carbon will combine the effects of filtration and adsorption. Therefore, for the sake of economical working, it will be preferable to remove the suspended solids as much as possible beforehand and leaving only the impurities in solution to be adsorbed by the activated carbon. This process will therefore be the ideal for the tertiary treatment of sewage.

3.3 Land

Land pollution is the degradation of the Earth's land surface through misuse of the soil by poor agricultural practices, mineral exploitation, industrial waste dumping, and indiscriminate disposal of urban wastes. It includes visible waste and litter as well as pollution of the soil itself.

Examples of Land Pollution

Soil Pollution

Soil pollution is mainly due to chemicals in herbicides (weed killers) and pesticides (poisons which kill insects and other invertebrate pests). Litter is waste material dumped in public places such as streets, parks, picnic areas, at bus stops and near shops.

Waste Disposal

The accumulation of waste threatens the health of people in residential areas. Waste decays, encourages household pests and turns urban areas into unsightly, dirty and unhealthy places to live in.

Control Measures

The following measures can be used to control land pollution:

- anti-litter campaigns can educate people against littering;
- organic waste can be dumped in places far from residential areas;
- inorganic materials such as metals, glass and plastic, but also paper, can be reclaimed and recycled

3.4 Marine Pollution

"Introduction of substances or energy into the marine environment (including estuaries) resulting in such deleterious effects as harm to living resources, hazard to human health, hindrance to marine activities including fishing, impairment of quality for use of sea-water, and reduction of amenities."

3.4.1 Sources and Effects of Marine Pollution Toxics

Toxic waste like Mercury and Toxic materials are substances derived from industrial, agricultural, household cleaning, gardening and automotive products are the most harmful form of pollution to marine creatures. Once a form of toxic waste affects an organism, it can be quickly passed along the food chain and might eventually end up as seafood, causing various problems. Toxic wastes arrive from the leakage of landfills, dumps, mines and farms. Sewage and industrial wastes introduce chemical pollutants like DDT and Sevin. Farm chemicals (insecticides and herbicides) along with heavy metals (e.g. mercury and zinc) can have disastrous effect on marine life.

Sewage and Fertilizers

The discharge of sewage can cause public health problems either from contact with polluted waters or from consumption of contaminated fish or shellfish. The discharge of untreated sewage effluents also produces long-term adverse impacts on the ecology of critical coastal ecosystems in localized areas due to the contribution of nutrients and other pollutants. Pollution due to inadequate sewage disposal causes nutrient enrichment around population centers, and high nutrient levels and even eutrophication near treatment facilities and sewage outfalls.

Around the world, untreated sewage flows into coastal waters, carrying organic waste and nutrients that can lead to oxygen depletion, as well as disease-causing bacteria and parasites that require closing beaches and shellfish beds (See Figure 4)

The inadequate number of sewage treatment plants in operation, combined with poor operating conditions of available treatment plants, and the disposal practices of discharging mostly untreated wastewater are likely to have an adverse effect on the ocean.

Oil

The sites most vulnerable for accidents are areas where tankers and barges move through restricted channels and in the vicinity of ports. In spite of regulations established, tankers and barges do not always use port facilities for the disposal of bilge and tank washing and wastes, and a significant amount of oil, which exceeds that from accidental oil spills is discharged into the coastal areas this way.

The impact of oil pollution on the ecology of coastal and marine ecosystems is particularly destructive following massive oil spills caused by maritime accidents. However, gas exchange between the water and the atmosphere is decreased by oil remaining on the surface of the water, with the possible result of oxygen depletion in enclosed bays where surface wave action is minimal. Coral death results from smothering when submerged oil directly adheres to coral surfaces, and oil slicks affect sea birds and other marine animals. In addition, tar accumulation on beaches reduces tourism potential of coastal areas.

Mining and Dredging

Mining is not actually a pollutant but it does affect the marine ecosystem and its habitat. Mining can erode beaches, degrade water quality, and spoil coastal habitats. Mining coral to process lime can remove the habitat of local marine species and weakens coastal storm defenses. Mined or dredged areas take a very long time to recover. Because of this, strict regulations govern the dredging of the ocean floor.

Synthetic Organic Chemicals

Many different synthetic organic chemicals enter the ocean and become incorporated into organisms. Ingestion of small amounts can cause illness or death. Halogenated hydrocarbons are a class of synthetic hydrocarbon compounds that contain chlorine, bromine, or iodine are used in pesticides, flame retardants, industrial solvents, and cleaning fluids.

The level of synthetic organic chemicals in seawater is usually very low, but some organisms can concentrate these toxic substances in their flesh at higher levels in the food chain. That is an example of biological amplification.

Marine Debris

More garbage such as plastic bags, rope, helium balloons, and stray fishing gear, build up in our oceans every year. Synthetic materials stay in the environment for years, killing or injuring ocean species, like right whales and leatherback turtles, which mistake litter for food or get entangled in it.

Plastic is a large amount and danger of solid waste. Plastic is not biodegradable and therefore affects the oceans for long periods of time. Sea turtles mistake plastic bags for jellyfish and die from internal blockages. Seals and sea lions starve after being muzzled by six-pack rings or entangled by nets.

3.4.2 Control Measures

The major source marine pollution is mainly due to spillage of oil and this can be controlled by the following method sans.

Chemical Methods

(i) Dispersants

Dispersants are chemicals, which have components of surface-active agents called surfactants (also known as detergents). The dispersants aids in the breaking up of the oil slick into smaller droplets. Chemical dispersants remove the oil from the surface of the water and into the water column by enhancing the natural chemical and physical breakdown of oil. Once in the water column, the dilution of oil is greatly facilitated and hence the toxicity level is reduced. Eventually, these dispersants and oil droplets are food sources for bacteria in the ocean. By removing the oil from the water surface, birds, marine mammals, turtles, and sensitive coasts are protected. Chemical dispersants do not cause the oil to sink, but remain in suspension in the water column.

The downside to chemical dispersants is that they are toxic in certain ways. As an added precaution, chemical dispersants are not applied to shallow near-shore waters, mangrove areas, marshes, or waters over coral reefs and sea grass beds.

(ii) Combustion

Through controlled burning of the oil can effectively remove the oil slick. This method is called in-situ burning. In-situ burning can approximately remove around 100-gallons/day/square foot of surface area under excellent weather conditions. In this way, bird, marine mammals, turtles and sensitive coast areas are being spared from the effects of the spill.

In-situ burning is only conducted when the winds are blowing away from or at a safe distance from the populated areas. The reason is that, during burning, tons of black smoke is being produced and this plume could travel miles from its source. The black smoke is mainly carbon dioxide and around 5-10 percent particulate with small amounts of carbon monoxide, nitrous oxide and sulphur dioxide. These are seen as major air pollutants and hence must be adequately contained through downwind dispersion and dilution.

However, there are some restrictions to in-situ burning. Only for sea levels of two to three feet or less, can in-situ burning be possible. Sea levels higher that that, in-situ burning would be discontinued. Another factor is the weather conditions. Only under, good weather conditions can this burning can be carried out.

Mechanical Means

(i) Booms

Booms may be used as a sweeping mode to concentrate the oil towards a skimmer so as to increase the oil encounter rate of that skimmer. In addition, booms are used to protect certain coastal resources from the effects of the oil spills or to deflect the oil on to a nominated or sacrificial beach for recovery purposes. In the former case, the booms are deployed from ships and used in a mobile mode. In the latter case, they may be moored and used in stationary modes.

During a spill response, sensitive locations threatened by an advancing oil slick can use booms to prevent any immediate adverse effects. Booms are floating barriers to oil, made of plastic, metal, or other materials. This prevents spreading, so that a smaller area is affected and clean up is easier. It also keeps oil from highly sensitive areas that are quickly damaged such as shellfish beds or, in this exercise, beaches used by piping plovers as nesting habitat.

(ii) Sorbents

This is done by collection of oil by means of oleophilic materials by adsorption in a variety of mechanical methods such as oil mopping, adsorption belts and adsorptive disc skimmers. The material used is often a small sized particle which may be broadcast on the oil to effect adsorption.

Generally speaking, adsorbents can be used only on a small scale and it should be remembered that the use of adsorbents increases the bulk of material, which must eventually be handled in the oil spill clearance operation.

Nevertheless within these limits they can be extremely useful and an enormous range of materials has already been used for these purposes.

(iii) Oil Skimmers

Once the oil is being enclosed and contained, it muse immediately be recovered. Skimmers mechanically remove oil from the water surface without causing major changes in its physical or chemical properties. Depending on the type of oil spills, the presence of debris, location of spill, ambient weather conditions and the calmness of the water the skimmers operate at different efficiencies.

Although designs vary, all oil skimmers rely on specific gravity; surface tension and a moving medium to remove floating oil from a fluid's surface. Floating or sinking oil and grease cling to skimming media more readily than water and water has little affinity for the media. This allows skimming media in the shape of a belt, disk, drum, etc. to pass through a fluid surface to pick up oil and grease with very little water. This oily material is subsequently removed from the media with wiper blades or pinch rollers.

Biological Means (Bioremediation)

Biological Methods have been developed, thereby microorganisms namely bacteria attached themselves to the sticky oil and dissolving it. This method will have a good scope for the future.

The bacteria eat up some of the oil and turn the remainder into a less viscous "oil-milk" of tiny oil droplets in the water. The more finely divided the oil is, the more accessible it becomes to other types of oil-eating bacteria. The great advantage of using bacteria to combat oil slicks is that there are no chemicals or surfactants such as detergents.

3.5 Noise

Noise pollution can be defined as unwanted or offensive sounds that unreasonably intrudes into our daily activities. It has many sources, most of which are associated with urban development: road, air and rail transport; industrial noise; neighborhood and recreational noise. A number of factors contribute to problems of high noise levels, including:

- increasing population, particularly where it leads to increasing urbanization and urban consolidation; activities associated with urban living generally lead to increased noise levels
- Increasing volumes of road, rail and air traffic.

3.5.1 Effects of Noise

Noise can affect human health and well-being in a number of ways, including annoyance reaction, sleep disturbance, interference with communication, performance effects, effects on social behavior and hearing loss. Noise can cause annoyance and frustration as a result of interference, interruption & distraction. Activity disturbance is regarded as an important indicator of the community impact of noise. The national noise survey assessed two major disturbances, for example, to listening activities and sleep: 41% of respondents reported experiencing disturbance to listening activities and 42% to sleep.

Research into the effects of noise on human health indicates a variety of health effects. People experiencing high noise levels (especially around airports or along road/ rail corridors) differ from those with less noise expo sure in terms of: increased number of headaches, greater susceptibility to minor accidents, increased reliance on sedatives and sleeping pills, increased mental hospital admission rates.

Exposure to noise is also associated with a range of possible physical effects including: colds, changes in blood pressure, other cardiovascular changes, increased general medical practice attendance, problems with the digestive system and general fatigue.

There is fairly consistent evidence that prolonged exposure to noise levels at or above 80 dB(A) can cause deafness. The amount of deafness depends upon the degree of exposure.

3.5.2 Major Noise Sources

Road Traffic

Road traffic noise is one of the most widespread and growing environmental problems in urban .The impact of road traffic noise on the community depends on various factors such as road location and design, land use planning measures, building design, vehicle standards and driver behavior.

Air Traffic

It is one of the cause of considerable community concern. The extent of aircraft noise impact depends on the types of aircraft flown, the number of flights and flight paths.

Rail Traffic

There are two main sources of noise and vibration relating to the operation of the rail network: the operation of trains and the maintenance and construction of rail infrastructure.

The level of noise associated with rail traffic is related to the type of engine or rolling stock used the speed of the train and track type and condition.

Industries

This is associated with the noise due to operation of plant machineries.

3.5.3 Noise Control Measures

Responsibility for Noise Control

The Environmental Planning and Assessment Act 1979 provides responsibility and opportunity for controlling environmental noise through the planning process. Consideration of the implications of environmental noise at the planning stage can often avoid or minimize the need for supplementary noise controls. However, in some instances noise reduction or mitigation measures are essential, for example:

- controls on noise levels generated from a source (e.g. vehicle/machine design, driver/operator behavior)
- controls on noise transmission (e.g. through the use of noise barriers)
- measures to reduce the level of sound reaching a receiver (e.g. soundproofing sensitive or affected buildings).

Reducing Road Traffic Noise

The Noise Control (Motor Vehicles and Motor Vehicle Accessories) Regulation 1995 prescribes noise levels for classes of motor vehicles and restricts allowable noise levels for vehicles manufactured at variable times depending on the class of the vehicle. In addition, the EPA conducts a noisy vehicle testing program on passenger cars, motor bikes and trucks. However, despite progress in addressing the problem of individually noisy vehicles, the rise in traffic volume has meant an increase in traffic noise overall.

Reducing Rail Traffic Noise

Under the provisions of the *Noise Control Act 1975* in NSW the railway system is classified as scheduled premises and as such the EPA has a regulatory role, and seeks to achieve noise targets for rail operations throughout the state to minimize the impact on local residents.

The railway sector has, in recent years, recorded an increase in the identification and reporting of noise problems by the community. There is a range of initiatives to address this issue, including:

- retrofitting existing locomotives to reduce noise emitted
- upgrading existing track to continuously welded rail which removes rail joints-a significant source of noise and vibration
- designing new bridges to reduce noise and retrofitting of existing bridges with noise attenuation devises
- deploying quieter rolling stock in noise sensitive areas
- use of electric locomotives wherever possible in the metropolitan and suburban area
- altering the holding pattern of trains to avoid them being held at signals for extended periods in built up areas.

Reducing Industrial Noise

The EPA issues licenses for the management of scheduled premises. When issuing a license the EPA sets initial noise limits that are achievable with the operation of plant and equipment currently installed, operated and maintained effectively. To achieve further improvements in noise exposure to residents, negotiations with the licensed premises are carried out and can be incorporated in the license as Pollution Reduction Programs (PRPs) . The EPA is currently working with industry to reduce noise levels from major sources.

Reducing Neighborhood Noise

The Noise Control (Miscellaneous Articles) Regulation 1995 was introduced to cover community noise issues not covered by previous legislation. It includes limitations on burglar alarms for both residential and commercial premises. Changes have been made to the night-time control of common domestic noise sources such as power tools, air conditioners, amplified music and lawn mowers. Under the new regulation only one warning to the offender is required and the warning is valid for 28 days. If an offence is committed within this period a fine can be issued without further warnings. The previous regulation warning was only active for 12 hours which meant it was not very effective with repetitious offences typical in suburban areas.

The Noise Control (Motor Vehicles and Motor Vehicle Accessories) Regulation 1995 controls the noise of individual motor vehicles. It includes a provision to control noise from a range of accessories including horns, alarms, refrigeration units and sound systems. It also places responsibility to ensure compliance of repairs/modifications of vehicles on the vehicle repairer.

The Noise Control (Marine Vessels) Regulation 1996 controls noise from boats in terms of the impact of noise to shoreline communities and residences. It includes controls on music on boats at night; controls on noisy exhausts; and a new vessels defect system to ensure all powered vessels operate to a standard.

Noise from barking dogs has been identified as a considerable problem in the country. Problems with dog noise are difficult to manage since control cannot be achieved by imposing noise level limits. Dog noise can be minimized by encouraging the owner to manage the dog in such a way that the barking behavior is modified. Control of dog noise is the responsibility of local councils.

Reducing Noise in Schools

In addition to the measures introduced to reduce the source and transmission of noise, measures can be undertaken to noise proof buildings thereby reducing the occupant exposure to noise.

A number of techniques are used to reduce noise levels in schools and emanating from schools, including:

- building finishes and construction details designed to minimize noise impacts in sensitive areas such as music rooms, technology workshops and gymnasiums
- a range of solutions to deal with road traffic noise, including acoustically sealing walls closest to the noise source, mechanically ventilating rooms and construction of barriers between the noise source and the affected site
- measures to deal with air traffic noise, including orientation of rooms to reduce noise exposure; use of noise attenuating materials and construction techniques; larger than normal roof overhangs and heavily insulated roofs with acoustically absorbent eaves linings; and soft finishes to the ground immediately adjacent to windows to reduce sound reflection into openings
- careful sitting of rooms housing noise producing activities (e.g. music rooms and workshops) to minimize impacts on neighbors beyond the school boundaries. This may also include careful location of windows.

Similar measures are used to minimize noise intrusion in other sensitive buildings such as hospitals. Design and building techniques can also be used to reduce traffic noise levels in the home.

3.6 Thermal Pollution

Industrial discharge of heated water into a river, lake, or other body of water, causing a rise in temperature that endangers aquatic life.

3.6.1 Effects of Thermal Pollution

Heat and hot water result from many industrial processes. They are in particular by-products of the activity of the power stations and nuclear. The water rejected into the marine mediums has harmful effects, primarily on the marine animal-life.

The pollution thermal has effects difficult to define with precision. One observes significant interferences of the reduction in the salinity of water and variations of this one in time and space. Several actions of such a pollution were however determined. It was in particular established that the hotter the temperature of rejected water is and more harmful is its effect on the benthic organizations

On the algae, the action of a thermal pollution is very variable. One noted opposite effects is strong proliferations or on the contrary a significant mortality. The metabolism of the algae, the phytoplanktons, the benthic macrophages is very faded by heated water.

3.7 Nuclear Hazards

An actual or potential release of radioactive material at a commercial nuclear power plant or a transportation accident or other radiological incident.

Sources

3.7.1 Nuclear Weapons

The effects of nuclear weapons are classified as either initial or residual. Initial effects occur in the immediate area of the explosion and are hazardous in the first minute after the explosion. Residual effects can last for days or years and cause death. The principal initial effects are blast and radiation.

Blast

Defined as the brief and rapid movement of air away from the explosion's center and the pressure accompanying this movement. Strong winds accompany the blast. Blast hurls debris and personnel, collapses lungs, ruptures eardrums, collapses structures and positions, and causes immediate death or injury with its crushing effect.

Thermal Radiation

The heat and light radiation a nuclear explosion's fireball emits. Light radiation consists of both visible light and ultraviolet and infrared light. Thermal radiation produces extensive fires, skin burns, and flash blindness.

Nuclear Radiation

Nuclear radiation breaks down into two categories-initial radiation and residual radiation.

Initial nuclear radiation consists of intense gamma rays and neutrons produced during the first minute after the explosion. This radiation causes extensive damage to cells throughout the body. Radiation damage may cause headaches, nausea, vomiting, diarrhea, and even death, depending on the radiation dose received. The major problem in protecting yourself against the initial radiation's effects is that you may have received a lethal or incapacitating dose before taking any protective action. Personnel exposed to lethal amounts of initial radiation may well have been killed or fatally injured by blast or thermal radiation.

Residual radiation consists of all radiation produced after one minute from the explosion. It has more effect on you than initial radiation. A discussion of residual radiation takes place in a subsequent paragraph.

3.7.2 Types of Nuclear Bursts

There are three types of nuclear bursts—airburst, surface burst, and subsurface burst. The type of burst directly affects your chances of survival. A subsurface burst occurs completely underground or underwater. Its effects remain beneath the surface or in the immediate area where the surface collapses into a crater over the burst's location. Subsurface bursts cause you little or no radioactive hazard unless you enter the immediate area of the crater. No further discussion of this type of burst will take place.

An airburst occurs in the air above its intended target. The airburst provides the maximum radiation effect on the target and is, therefore, most dangerous to you in terms of immediate nuclear effects.

A surface burst occurs on the ground or water surface. Large amounts of fallout result, with serious long-term effects for you. This type of burst is your greatest nuclear hazard.

3.7.3 Nuclear Injuries

Most injuries in the nuclear environment result from the initial nuclear effects of the detonation. These injuries are classed as blast, thermal, or radiation injuries. Further radiation injuries may occur if you do not take proper precautions against fallout. Individuals in the area near a nuclear explosion will probably suffer a combination of all three types of injuries.

Blast Injuries

Blast injuries produced by nuclear weapons are similar to those caused by conventional high-explosive weapons. Blast overpressure can produce collapsed lungs and ruptured internal organs. Projectile wounds occur as the explosion's force hurls debris at you. Large pieces of debris striking you will cause fractured limbs or massive internal injuries. Blast over-pressure may throw you long distances, and you will suffer severe injury upon impact with the ground or other objects. Substantial cover and distance from the explosion are the best protection against blast injury. Cover blast injury wounds as soon as possible to prevent the entry of radioactive dust particles.

Thermal Injuries

The heat and light the nuclear fireball emits causes thermal injuries. First-, second-, or third-degree burns may result. Flash blindness also occurs. This blindness may be permanent or temporary depending on the degree of exposure of the eyes. Substantial cover and distance from the explosion can prevent thermal injuries. Clothing will provide significant protection against thermal injuries. Cover as much exposed skin as possible before a nuclear explosion. First aid for thermal injuries is the same as first aid for burns. Cover open burns (second-or third-degree) to prevent the entry of radioactive particles. Wash all burns before covering.

Radiation Injuries

Neutrons, gamma radiation, alpha radiation, and beta radiation cause radiation injuries. Neutrons are high-speed, extremely penetrating particles that actually smash cells within your body. Gamma radiation is similar to X rays and is also a highly penetrating radiation. During the initial fireball stage of a nuclear detonation, initial gamma radiation and neutrons are the most serious threat. Beta and alpha radiation are radioactive particles normally associated with radioactive dust from fallout. They are short-range particles and you can easily protect yourself against them if you take precautions. See Bodily Reactions to Radiation, below, for the symptoms of radiation injuries.

3.7.4 Residual Radiation

Residual radiation is all radiation emitted after 1 minute from the instant of the nuclear explosion. Residual radiation consists of induced radiation and fallout.

Induced Radiation

It describes a relatively small, intensely radioactive area directly underneath the nuclear weapon's fireball. The irradiated earth in this area will remain highly radioactive for an extremely long time. You should not travel into an area of induced radiation.

Fallout

Fallout consists of radioactive soil and water particles, as well as weapon fragments. During a surface detonation, or if an airburst's nuclear fireball touches the ground, large amounts of soil and water are vaporized along with the bomb's fragments, and forced upward to altitudes of 25,000 meters or more. When these vaporized contents cool, they can form more than 200 different radioactive products. The vaporized bomb contents condense into tiny radioactive particles that the wind carries and they fall back to earth as radioactive dust. Fallout particles emit alpha, beta, and gamma radiation. Alpha and beta radiation are relatively easy to counteract, and residual gamma radiation is much less intense than the gamma radiation emitted during the first minute after the explosion. Fallout is your most significant radiation hazard, provided you have not received a lethal radiation dose from the initial radiation.

3.8 Nuclear Waste and their Teratment

Nuclear waste can be diveded into natural and man-made sources

Natural

Natural source may be cosmic radiation from space, or naturall occurring isotopes in the Earth's crust, For exmaple Coal contains a small amount of radioactive nuclides, such as uranium and thorium, but it is less than the average concentration of those elements in the Earth's crust. They become more concentrated in the fly ash because they do not burn well. However, the radioactivity of fly ash is still very low. It is about the same as black shale and is less than phosphate rocks, but is more of a concern because a small amount of the fly ash ends up in the atmosphere where it can be inhaled.

Similarly Residues from the oil and gas industry often contain radium and its daughters. The sulphate scale from an oil well can be very radium rich, while the water, oil and gas from a well often contains radon. The radon decays to form solid radioisotopes which form coatings on the inside of pipework. In an oil processing plant the area of the plant where propane is processed is often one of the more contaminated areas of the plant as radon has a similar boiling point as propane.

Man-made Sources

Waste from these sources arising from mineral processing of radiocative elements, medical, nuclear explosion and atomic power plant.

Mineral Processing

Wastes from mineral processing can contain natural radioactivity.

Medical

Radioactive medical waste tends to contain beta ray and gamma ray emitters. It can be divided into two main classes. In diagnostic nuclear medicine a number of short-lived gamma emitters such as ^{99m}Tc are used. Many of these can be disposed of by leaving it to decay for a short time before disposal as normal trash. Other isotopes used in medicine, with half-lives in parentheses:

- ^{90}Y, used for treating lymphoma (2.7 days)
- ^{131}I, used for thyroid function tests and for treating thyroid cancer (8.0 days)
- ^{89}Sr, used for treating bone cancer, intravenous injection (52 days)
- ^{192}Ir, used for brachytherapy (74 days)
- ^{60}Co, used for brachytherapy and external radiotherapy (5.3 years)
- ^{137}Cs, used for brachytherapy, external radiotherapy (30 years)

Industrial

Spent nuclear reactor fuel from both commercial power plants and military facilities, as well as reprocessed materials which can emit large amounts of radiation for hundreds of thousands of years.

Mill tailings, left over when ore is refined and processed is the largest by volume of any form of radioactive waste. Only 1% of uranium ore contains uranium—the rest is left on-site as sandlike residue. These tailings are generally left outdoors in huge piles, where they blow around, releasing radioactive materials into the surrounding air and water. By 1989, some 140 million tons of mill tailings had accumulated in the United States alone, with 10 to 15 million tons added each year. Although their radiation is generally less concentrated than other types of waste, some of the isotopes in these tailings are long-lived and can be hazardous for many thousands of years.

Industrial source waste can contain alpha beta, neutron or gamma emitters. Gamma emitters are used in radiography while neutron emitting sources are used in a range of applications, such as oil well logging.

Nuclear Weapons Production

Waste from nuclear weapons reprocessing (as opposed to production, which requires primary processing from reactor fuel) is unlikely to contain much beta or gamma activity other than tritium. It is more likely to contain alpha emitting actinides such as ^{239}Pu which is a fissile material used in bombs, plus some material with much higher specific activities, such as ^{238}Pu or Po. Some of these actinides have been used in neutron triggers for bombs (although this is not the preferred present method) or for thermoelectric power sources such as the ^{238}Pu used in outer plantetary probes like Voyager, Galileo, and Cassini.

Types of Radioactive Waste

There are three types of Nuclear waste namely Low level, Medium level and High level waste,

Low Level Waste

These waste is usually orginating from from hospitals and industry, as well as the nuclear fuel cycle. It comprises paper, rags, tools, clothing, filters, etc., which contain small amounts of mostly short-lived radioactivity. Commonly,

LLW waste is designated as such as a precautionary measure if it originated from any region of an 'Active Area', which frequently includes offices with only a remote possibility of being contaminated with radioactive materials. Such LLW waste typically exhibits no higher radioactivity than one would expect from the same material disposed of in a non-active area, such as a normal office block. No LLW waste requires shielding during handling and transport and is suitable for shallow land burial. To reduce its volume, it is often compacted or incinerated before disposal. Low level waste is divided into four classes, class A, B, C and GTCC, which means greater than class C.

Medium Level Waste (MLW) contains higher amounts of radioactivity and in some cases requires shielding. MLW includes resins, chemical sludge and metal fuel cladding, as well as contaminated materials from reactor decommissioning.

High Level Waste (HLW) arises from the use of uranium fuel in a nuclear reactor and nuclear weapons processing. It contains fission products and transuranic elements generated in the reactor core. It is highly radioactive and hot. HLW accounts for over 95% of the total radioactivity produced in the process of nuclear electricity generation.

Treatment of Medium Level Waste

It is common for medium active wastes in the nuclear industry to be treated with ion exchange or other means to concentrate the radioactivity into a small volume. The much less radioactive bulk (after treatment) is often then discharged. For instance it is possible to use a ferric hydroxide floc to remove radioactive metals from aqueous mixtures. After the radioisotopes are absorbed onto the ferric hydroxide the resulting sludge can be placed in a metal drum before being mixed with cement to form a solid waste form. In order to get better long term performance instead of *normal* cement a mixture of fly ash or blast furnace slag and portland cement can be used.

Treatment of High Level Waste

Storage

In the absence of high-level waste repositories, nuclear power plants generally store their spent fuel rods in lead-lined concrete pools of water. These pools somewhat contain the spread of gamma radiation by keeping the rods relatively cool. They also help prevent fission.

Stabilisation or Vitrification

Long-term storage of radioactive waste requires the stabilization of the waste into a form which will not react, nor degrade, for extended periods of time. Vitrification is the process in which the high-level waste is mixed with sugar and calcined. Calcination involves passing the waste through a heated, rotating tube to evaporate the water off the waste, and de-nitrate the fission products to assist the stability of the glass produced. The 'Calcine' generated is fed continuously into a induction heated furnace with fragmented glass. The resulting glass is a new compound where the waste products are bonded into the glass' matrix. This is poured into stainless steel containers in a batch process. Thus, when cooled the waste products are immobilised for a very long period of time (many thousands of years). After filling, a seal is welded onto the cylinder which is then washed and stored after inspection.After being inspected for external contamination, the steel cylinder is placed in a store. Various types galss like borosilicate glass, phosphate glass, are used stabilisation.

Synroc

Zircon ($ZrSiO_4$) and more complex ceramics such as, e.g., pyrochlore, perovskite, and zirconolite, which are part of the Synroc ceramic waste form, have been used widely to safely encapsulate highly radioactive waste. Zircon is also widely used in geochronology and has recently become a model 'host' matrix for studying encapsulation of plutonium waste in ceramic materials .The synroc contains pyrochlore and cryptomelane type minerals. The original form of synroc (synroc C) was designed for the liquid high level waste (PUREX raffinate) from a ligh water reactor. The main minerals in this synroc are hollandite ($BaAl_2Ti_6O_{16}$), zirconolite ($CaZrTi_2O_7$) and perovskite ($CaTiO_3$). The zirconolite and perovskite are hosts for the actinides. The strontium and barium will be fixed in the perovskite. The cesium will be fixed in the hollandite.

Geological Disposal

For long-term storage of high-level waste, a waterproof, geologically stable repository and leak-proof waste container is required. Packaging has to be tailored to the volume of the waste, the actual radioactive **isotopes** of elements it contains, how radioactive it is, its isotopes' half-lives, and how much heat it still generates. One technique for packaging high-level wastes involves melting them with glass and pouring the molten material into impermeable containers. The containers could be buried in soil or in a rock pile and surrounded by fill material and a barrier wall. From the 1940s through the 1960s, barrels of radioactive waste were frequently dumped in oceans. In offshore drilling, holes would be drilled into the seabed and filled with barrels of waste. In self-burial, specially shaped barrels would be dumped and left to sink to the ocean floor.

Sea-based options for disposal of radioactive waste include burial beneath a stable abyssal plain, burial in a subduction zone that would slowly carry the waste downward into the Earth's mantle, and burial beneath a remote natural or human-made island. While these approaches all have merit and would facilitate an international solution to the vexing problem of disposal of radioactive waste, they are currently not being seriously considered because of the legal barrier of the Law of the Sea.

Reuse of Waste

Another choice is to find applications of the isotopes in nuclear waste so as to reuse them. Already, cesium 137, strontium 90, technetium 99, and a few other isotopes are extracted for certain industrial applications like food irradiation.

3.9 Solid Waste Management

Solid Waste Disposal

Disposal of normally solid or semisolid materials, resulting from human and animal activities, that are useless, unwanted, or hazardous. Solid wastes typically may be classified as follows:

Garbage: decomposable wastes from food

Rubbish: non-decomposable wastes, either combustible (such as paper, wood, and cloth) or noncombustible (such as metal, glass, and ceramics)

Ashes: residues of the combustion of solid fuels.

*Largewastes:*demolition and construction debris and trees and Dead animals.

Sewage-treatment solids: material retained on sewage-treatment screens, settled solids, and biomass sludge.

Industrial wastes: such materials as chemicals, paints, and sand.

Mining wastes: slag heaps and coal refuses piles.

Agricultural wastes: farm animal manure and crop residues.

Disposal of solid wastes on land is by far the most common method in many countries and probably accounts for more than 90 percent of the nation's municipal refuse. Incineration accounts for most of the remainder, whereas composting of solid wastes accounts for only an insignificant amount. Selecting a disposal method depends almost entirely on costs, which in turn are likely to reflect local circumstances.

(i) Landfill

Sanitary landfill is the cheapest satisfactory means of disposal, but only if suitable land is within economic range of the source of the wastes; typically, collection and transportation account for 75 percent of the total cost of solid waste management. In a modern landfill, refuse is spread in thin layers, each of which is compacted by a bulldozer before the next is spread. When about 3 m (about 10 ft) of refuse has been laid down, it is covered by a thin layer of clean earth, which also is compacted.

Pollution of surface and groundwater is minimized by lining and contouring the fill, compacting and planting the cover, selecting proper soil, diverting upland drainage, and placing wastes in sites not subject to flooding or high groundwater level.

(ii) Incinerators

In incinerators of conventional design, refuse is burned on moving grates in refractory-lined chambers; combustible gases and the solids they carry are burned in secondary chambers. Combustion is 85 to 90 percent complete for the combustible materials. In addition to heat, the products of incineration include the normal primary products of combustion is carbon dioxide and water as well as oxides of sulfur and nitrogen and other gaseous pollutants; nongaseous products are fly ash and unburned solid residue. Emissions of fly ash and other particles are often controlled by wet scrubbers, electrostatic precipitators, and bag.

(iii) Composting

Composting operations of solid wastes include preparing refuse and degrading organic matter by aerobic microorganisms. Refuse is presorted, to remove materials that might have salvage value or cannot be composted, and is ground up to improve the efficiency of the decomposition process. The refuse is placed in long piles on the ground or deposited in mechanical systems, where it is degraded biologically to a humus with a total nitrogen, phosphorus, and potassium content of 1 to 3 percent, depending on the material being composted. After about three weeks, the product is ready for curing, blending with additives, bagging, and marketing.

Composting and Organic Waste

Waste from the garden, yard, and table does not have to be thrown away. It may be condensed and reused as a fertilizer through a process called composting. A compost pile may be built by layering different kinds of waste in a bin, leaving space between the layers for air to circulate. Nitrogen is added to the pile in the form of manure, meal, or greenery to generate heat. Heat facilitates rotting and kills all undesirable organisms. Once the pile is slightly dampened, it is covered. As heat and steam build up, the waste decomposes over time into a nutrient-rich substance called compost. The compost is then applied to plants as a fertilizer.

(iv) Resource Recovery

Numerous thermal processes, now in various stages of development, recover energy in one form or another from solid waste. These systems fall into two groups: combustion processes and pyralysis processes. A number of companies burn in-plant wastes in conventional incinerators to produce steam. A few municipalities produce steam in incinerators in which the walls of the combustion chamber are lined with boiler tubes; the water circulated through the tubes absorbs heat generated in the combustion chamber and produces steam.

Pyrolysis, also called destructive distillation, is the process of chemically decomposing solid wastes by heat in an oxygen-reduced atmosphere. This results in a gas stream containing primarily hydrogen, methane, carbon monoxide, carbon dioxide, and various other gases and inert ash, depending on the organic characteristics of the material being pyrolyzed.

(v) Recycling

The practice of recycling solid waste is an ancient one. Metal implements were melted down and recast in prehistoric times. Now a days, recyclable materials are recovered from municipal refuse by a number of methods, including shredding, magnetic separation of metals, air classification that separates light and heavy fractions, screening, and washing. Another method of recovery is the wet pulping process: Incoming refuse is mixed with water and ground into a slurry in the wet pulper, which resembles a large kitchen disposal unit. Large pieces of metal and other non-pulpable materials are pulled out by a magnetic device before the slurry from the pulper is loaded into a centrifuge called a liquid cyclone. Here the heavier non-combustibles, such as glass, metals, and ceramics, are separated out and sent on to a glass- and metal-recovery system; other, lighter materials go to a paper-fiber-recovery system. The final residue is either incinerated or is used as landfill.

Increasingly, municipalities and private refuse-collection organizations are requiring those who generate solid waste to keep bottles, cans, newspapers, cardboard, and other recyclable items separate from other waste. Special trucks pick up this waste and cart it to transfer stations or directly to recycling facilities, thus lessening the load at incinerators and land.

3.9.1 Role of Human in Prevention of Pollution

1. To increase awareness of the effects of waste and hazardous materials form home and business activities on the environment, and ways to prevent or reduce waste and hazardous materials problems.

   Public discourse and Training can be conducted and by this training

   Citizens will receive information and training in handling, storing and disposing of wastes and hazardous materials.

   Citizens will receive training and information on how to implement home- or commercial-scale composting.

   Citizens wil receive training and information on assessing the risks of hazardous conditions in and around the home and farm.

2. To increase awareness among citizens of the impacts of water use on water suply and water quality.

   Public discourse and Training can be conducted by this training

   Citizens will receive training and information on water conservation practices for lawns and gardens.

   Citizens will receive training and information in protecting water supply fro non-point source pollution (NPS).

3. To increase awareness, understanding and implementation of safe pesticide practices in homes, institutions, growing operations and businesses.

   Public discourse and Training can be conducted and by this training

   Citizens will receive training and information in proper selection, storage and use of pesticides.

   Citizens will receive training and information in non-chemical pest-control techniques.

   Citizens will receive training and information in the state certification porgram for private, commercial and institutional pesticide applicators.

4. To increase awareness, understanding and preservation or development of wildlife habitat on public and private lands.

   Public discourse and Training can be conducted. By this training

   Citizens will receive training and information in wildlife habitat establishment, improvement and maintenance.

   Citizens will receive training and information on the importance of wildlife habitat to environmental quality and quality-of-life.

5. To increase awareness, understanding and information among all citizens regarding environmental quality issues (air, land and water), Public discourse and Training can be conducted.

3.10 Disaster Management

3.10.1 Floods

" Flooding is generally defined as any abnormally high stream flow that overtops the natural or artificial banks of a stream. Flooding is a natural characteristic of rivers. The floodplains are normally dry land areas. They are an integral part of a river system that acts as a natural reservoir and temporary channel for flood waters. If more runoff is generated than the banks of a stream channel can accommodate, the water will overtop the stream banks and spread over the floodplain. The ultimate factor of damage, however, is not the quantity of water being discharged but how high the water goes above normal restraints or embankments. Furthermore, floods can form where there is no stream, as for example when abnormally heavy precipitation falls on flat terrain at such a rate that the soil cannot absorb the water or the water cannot run off as fast as it falls.

Of all the disasters, flood disasters affect the most people. But there are many more flood disasters than droughts, and the number affected by floods is increasing much more rapidly than those suffering droughts. In fact, flooding is one natural hazard that is becoming a greater threat rather than a constant or declining one. Floods are caused not only by rain but also by human changes to the surface of the earth. Farming, deforestation, and urbanization increase the runoff from rains; thus storms that previously would have caused no flooding today inundate vast areas.

Not only do we contribute to the causes of floods, but reckless building in vulnerable areas, poor watershed management, and failure to control the flooding also help create the disaster condition.

Historical Examples of Bangladesh, 1974

Bangladesh is a riverine country where recurrent flooding is both common and necessary. Every year large areas are submerged during the monsoon season and fertilized by deposits of fresh alluvium, i.e., the soil deposited by moving water. However, if the waters remain stagnant for too long, these beneficial floods become major disasters. Such was the case in the summer and fall of 1974 when flooding extended over nearly one-half of the country and stagnated for more than a month. At least 1,200 people died in the floods and another 27,500 died from subsequent disease and starvation. Approximately 425,000 houses were destroyed or severely damaged and the losses to agriculture were estimated at U.S. $325.9 million. A total of 36 million people suffered severe hardship and losses due to the disaster.

There is a wide range of measures that can be used to protect against flooding. They may be grouped in various ways, such as:

Structural" and "nonstructural" measures.

Whether they are most suitable for protecting: a) individual structures or b) areas containing multiple structures and communities.

Whether their purpose is to: a) modify the flood; b) reduce susceptibility to flooding; and/or c) reduce the impact of flooding.

Multiple measures are usually needed to provide protection to an area.

The National Weather reporting department is responsible for warning the public of the possibility of flooding. Flood predictions generally are made at the regional "Weather Forecaste Center". There are several different warning messages that may be issued, based upon the conditions and/or probability of flooding.

3.10.2 Earthquake

An earthquake, like other disasters, has the potential to affect the lives of thousands in a community and the surrounding region While structural damage is the most visible effect of an earthquake, a key point to remember is that earthquake effects go beyond structural damage.

Seismic hazards are effects of an earthquake that may affect the normal activities of people.

This includes:

- Surface faulting.
- Ground shaking.
- Landslides.
- Liquefaction, where water-saturated ground acts as a fluid.
- Tectonic (rock) deformation.
- Subsidence.
- Lateral spreading.
- Tsunamis.
- Seiches, movement of a closed body of water, such as a lake, swimming pool, or fuel storage tank.
- Volcanoes.
- Delta failure.

Emergency management is an organized, four-phase process by which communities:

- **Prepare** for hazards that cannot be prevented, or mitigated.
- **Respond** to emergencies that occur.
- **Recover** from emergencies to restore the community to its pre-emergency condition.
- **Mitigate** risks by reducing or eliminating damage and disruption from future disasters.

3.10.3 Cyclone

Cyclones often produce winds in excess of 200 km/h which can cause extensive damage to property and turn debris into dangerous missiles Cyclones can also bring flooding rains, which cause further damage to property, and increase the risk of drowning Cyclones can cause huge seas, putting vessels in danger both in harbour and out at sea Most deaths from cyclones occur as a result of drowning, collapsed buildings, or flying debris which becomes lethal in high winds.

Tropical cyclones (also known as hurricanes in North America and typhoons in Asia) are giant whirlwinds of air and dense cloud spiraling at over 120 km/hr around a central 'eye' of extreme low pressure.

Australia's cyclone season is usually from December to April and affects most of the Queensland coast. The greatest threat lies north of the Tropic of Capricorn.

When a tropical depression develops and its associated winds reach gale force, it will be classified as a tropical cyclone and will be given a name.

Cyclones occur frequently in the Southern Hemisphere with an average of 10 cyclones per year tracked by the Bureau of Meteorology in the Australian region.

The 'life-cycle' of the average tropical cyclone is about seven days, but can extend to over three weeks.

Cyclone Watch

A cyclone watch is issued by the Bureau of Meteorology when a cyclone or developing cyclone is likely to affect coastal or inland communities within 24 to 48 hours.

A cyclone watch will include an estimate of the cyclone's position, its intensity, severity and movement.

Cyclone watches will be issued every three hours initially and hourly once the cyclone nears the coast.

Cyclone Warnings

A cyclone warning is issued by the Bureau of Meteorology when a cyclone or developing cyclone is likely to affect coastal or inland communities within 24 hours.

Warnings will identify the communities likely to be affected, the name of the cyclone, its position, intensity, severity and movement.

Communities under threat will be advised to take certain precautions to safeguard life and property.

3.10.4 Landslides

Landslides usually involve the movement of large amounts of either earth, rock, sand or mud or any combination of these.

Landslides can be caused by earthquakes, volcanoes, soil saturation from rainfall or seepage or by human activity (eg. vegetation removal, construction on steep terrain).

The rate of movement of a landslide can vary from exceptionally slow - centimeters per year, to a sudden and total collapse - such as an avalanche of perhaps millions of tonnes of debris.

The distance traveled by landslide debris can also vary greatly, from a few centimeters in 'ground slumps', to many kilometers when large mud flows follow river valleys.

Prevention

Prevention includes the identification of hazards, the assessment of threats to life and property, and the taking of measures to reduce potential loss of life and property damage, sometimes known as disaster mitigation.

Mitigation measures range from community awareness campaigns to increase knowledge of how to deal with disaster situations, land use planning and design decisions to stop development which may be dangerous in the event of a disaster, to capital works such as levee bank construction to reduce the impacts of flooding. All mitigation measures are important as they can not only reduce the cost of disasters to the community, but they help save lives.

Disaster Managers at all levels are responsible for using a risk management process to identify prevention and mitigation options.

Preparedness

Preparedness includes arrangements or plans to deal with a disaster or the effects of a disaster. Disaster District and Local Government Plans are developed to provide for the activation of the Disaster Management System and provision of resources to be used in case of a disaster.

Test Your Understanding

1. Explain the basics of Air pollution and its effects
2. Discuss the causes of water pollution
3. What are the various pollutants present in the waste water? Explain
4. Discuss primary treatment of waste water
5. What is activated sludge treatment system? Explain
6. Discuss various tertiary treatment process
7. Explain the sources and effects of Marine pollution
8. What is thermal Pollution?. And Explain its effects

9. Discuss the various sources of Nuclear hazards'
10. Explain various sources of Nuclear waste and its treatment
11. What are various types of solid waste? and discuss their treatment

    Explain the role human in preventing of pollution

CHAPTER 4

Social Issues and the Environment

4.1 Urban Issues

(i) Water Problem

The world's population is increasingly found in the cities. Now a days, throughout the developing world, urbanization trends are gaining speed and are irreversible. From a technical standpoint, it is easier to provide water and sanitation services to people living closer together in urban settings than in dispersed rural communities. However, the costs of meeting the needs are much higher per capita, and are growing. The health risks posed by the lack of sanitation increase exponentially as densities increase and as people share drinking water and sanitation resources.

The urban environmental sanitation crisis in developing countries is taking a large health, economic, and environmental toll on all city residents. Willingness to pay for basic water and sanitation services is often high in peri-urban neighbourhoods, provided that services are appropriate, effective, and affordable. The use of a strategic sanitation approach should helps to build capacity within implementing agencies and enhances the ability of communities to make sustainable sanitation improvements.

(ii) Energy Problem

Energy is basic to development. Modern energy services are powerful engine of economic and social opportunity: no country has managed to develop much beyond a subsistence economy without ensuring at least minimum access to energy services for a broad section of its population. It is not surprising to find, therefore, that the billion who live in developing countries attach a high priority to energy services. On average, these people spend nearly 12% of their income on energy. At the same time, the provision of energy services especially through the combustion of fossil fuels and biomass can create adverse environmental effects. In rich countries, much attention is directed to the regional and global consequences of fuel combustion, because many of the local effects have been controlled at considerable expense over the past half-century. In developing countries, the local environmental problems associated with energy use remain matters of concern that are as, or even more, urgent than they were in industrialized countries 50 or 100 years ago. Further, it is the poor who suffer most severely from such problems, because it is they who are forced to rely upon the most inefficient and polluting sources of energy services for lack of access to better alternatives.

4.2 Environmental Ethics-Issues and Possible Solutions

The formation of an ethics of the environment has taken place within the past twenty years, even though the problems of judging human dealings with respect to the environment and in terms of morality have been known for a longer time. The past twenty years, however, have seen the recognition of the ethics of the environment as a scientific discipline, an institutionalization. Most of all this meant an intensive development, especially in the area of its categorical apparatus and the formation of its philosophical and methodological alternatives.

In our context the mode of forming and processing problems of environmental ethics begins rather late. This has a bearing mainly on our past political development, when the ethics of the protection of the environment were subordinated to the aims and intentions of the ruling power. The protection of the environment was pushed to the edge of interests while other values, especially those of social concerns, were focused upon. Emphasis was placed on the construction of mass settlements, the building of factories and the like, but without much consideration being given to the impact of such decisions upon the environment. The problematic of the environment consequently remained rather in the mind of the political opposition

In the recent times exploration in that field has noticeably improved. Proposals for environmental laws1 are being formulated, tens of non-governmental organizations for the protection of the environment were created, the formation of institutional assumptions for environmental ethics as a science is carried out, and many articles and studies focused on the problematic of the protection of the environment are being published. Regularly every year seminars and conferences on that theme are being organized; The time has come when it will be necessary to include this problematic purposefully even into training and educational programmes. Today's opinions concerning environmental ethics are pluralistic and broadly spread.

4.3 Water Conservation

Our ancient religious texts and epics give a good insight into the water storage and conservation systems that prevailed in those days. Over the years rising populations, growing industrialization, and expanding agriculture have pushed up the demand for water. Efforts have been made to collect water by building dams and reservoirs and digging wells; some countries have also tried to recycle and desalinate (remove salts) water. Water conservation has become the need of the day. The idea of ground water recharging by harvesting rainwater is gaining importance in many cities.

In the forests, water seeps gently into the ground as vegetation breaks the fall. This groundwater in turn feeds wells, lakes, and rivers. Protecting forests means protecting water 'catchments'. In ancient India, people believed that forests were the 'mothers' of rivers and worshipped the sources of these water bodies.

Some Ancient Indian Methods of Water Conservation

The Indus Valley Civilization, that flourished along the banks of the river Indus and other parts of western and northern India about 5,000 years ago, had one of the most sophisticated urban water supply and sewage systems in the world. The fact that the people were well acquainted with hygiene can be seen from the covered drains running beneath the streets of the ruins at both Mohenjodaro and Harappa. Another very good example is the well-planned city of Dholavira, on Khadir Belt, a low plateau in the Rann in Gujarat. One of the oldest water harvesting systems is found about 130 km from Pune along Naneghat in the Western Ghats. A large number of tanks were cut in the rocks to provide drinking water to tradesmen who used to travel along this ancient trade route. Each fort in the area had its own water harvesting and storage system in the form of rock-cut cisterns, ponds, tanks and wells that are still in use today. A large number of forts like Raigad had tanks that supplied water.

In ancient times, houses in parts of western Rajasthan were built so that each had a rooftop water harvesting system. Rainwater from these rooftops was directed into underground tanks. This system can be seen even today in all the forts, palaces and houses of the region.

Underground baked earthen pipes and tunnels to maintain the flow of water and to transport it to distant places, are still functional at Burhanpur in Madhya Pradesh, Golkunda and Bijapur in Karnataka, and Aurangabad in Maharashtra.

4.3.1 Rainwater Harvesting

In urban areas, the construction of houses, footpaths and roads has left little exposed earth for water to soak in. In many parts of the rural areas, flood water quickly flows to the rivers, which then dry up soon after the rains stop. If this water made to be absorbed in to ground and the groundwater supply get recharged. Rainwater harvesting essentially means collecting rainwater on the roofs of building and storing it underground for later use. Not only does this recharging arrest groundwater depletion, it also raises the declining water table and can help augment water supply. Rainwater harvesting and artificial recharging are picking up now a days thereby sea-water ingress is being arrested, and surface water run-off during the rainy season is being conserved.

The method is collect the rain water collected on rooftops and other surfaces, and the water is carried down to where it can be used immediately or stored or otherwise it can be directed to plants, trees or lawns or even to the aquifer.

Some of the benefits of rainwater harvesting are as follows

Increases water availability

Checks the declining water table

Is environmentally friendly

Improves the quality of groundwater through the dilution of fluoride, nitrate, and salinity

Prevents soil erosion and flooding especially in urban areas

4.4 Greenhouse Gases and their Sources

Carbon dioxide is one of the most important greenhouse gas in the atmosphere. Changes in land use pattern, deforestation, land clearing, agriculture, and other activities have all led to a rise in the emission of carbon dioxide. Methane is another important greenhouse gas in the atmosphere. About ¼ of all methane emissions are said to come from domesticated animals such as dairy cows, goats, pigs, buffaloes, camels, horses, and sheep. Domestic animals produce methane during the cud-chewing process. Methane is also released from rice or paddy fields that are flooded during the sowing and maturing periods. When soil is covered with water it becomes anaerobic or lacking in oxygen. Under such conditions, methane-producing bacteria and other organisms decompose organic matter in the soil to form methane. Nearly 90% of the paddy-growing area in the world is found in Asia, as rice is the staple food there. China and India, between them, have 80-90% of the world's rice-growing areas.

Methane is also emitted from landfills and other waste dumps. If the waste is put into an incinerator or burnt in the open, carbon dioxide is emitted. Methane is also emitted during the process of oil drilling, coal mining and also from leaking gas pipelines (due to accidents and poor maintenance of sites).

A large amount of nitrous oxide emission has been attributed to fertilizer application. This in turn depends on the type of fertilizer that is used, how and when it is used and the methods of tilling that are followed. Contributions are also made by leguminous plants, such as beans and pulses that add nitrogen to the soil.

4.4.1 Climate Change

Climate is changing day by day due to the following reasons.

The Industrial Revolution in the 19th century saw the large-scale use of fossil fuels for industrial activities. These industries created jobs and over the years, people moved from rural areas to the cities. This trend is continuing even today. More and more land that was covered with vegetation has been cleared to make way for houses. Natural resources are being used extensively for construction, industries, transport, and consumption. Consumerism (our increasing want for material things) has increased by leaps and bounds, creating mountains of waste. Also, our population has increased to an incredible extent.

All this has contributed to a rise in greenhouse gases in the atmosphere. Fossil fuels such as oil, coal and natural gas supply most of the energy needed to run vehicles, generate electricity for industries, households, etc. The energy sector is responsible for about ¾ of the carbon dioxide emissions, 1/5 of the methane emissions and a large quantity of nitrous oxide. It also produces nitrogen oxides (NOx) and carbon monoxide (CO) which are not greenhouse gases but do have an influence on the chemical cycles in the atmosphere that produce or destroy greenhouse gases.

4.4.2 Global Warming

Global warming is an observed increase in the average temperature of the Earth's atmosphere and oceans. Part of this increase may be due to natural processes, and would have occurred independently of human activity. The remainder is due to a human-induced intensification of the greenhouse effect. The increased volumes of carbon dioxide and other greenhouse gases released by the burning of fossil fuels, land clearing and agriculture, and other human activities, are the primary sources of human-induced warming. The natural greenhouse effect keeps the Earth about 33°C warmer than it otherwise would be; adding carbon dioxide to an atmosphere, with no other changes, will make a planet's surface warmer.

Since the beginning of the industrial revolution, atmospheric concentrations of carbon dioxide have increased nearly 30%, methane concentrations have more than doubled, and nitrous oxide concentrations have risen by about 15%. These increases have enhanced the heat-trapping capability of the earth's atmosphere. Sulfate aerosols, a common air pollutant, cool the atmosphere by reflecting light back into space; however, sulfates are short-lived in the atmosphere and vary regionally.

4.4.3 The Greenhouse Effect

Electromagnetic radiation is an exceedingly important physical phenomenon that takes various forms depending on its wavelength. Ordinary radios use the longest wavelengths of interest, 200 to 600 meters for AM stations. FM radio and television use wavelengths from a few meters (UHF) to less than 1 meter (VHF). Microwaves, familiar from the ovens that employ them, and radar, which plays a vital role in military applications and is also used by the police to catch speeders, have shorter wavelengths, in the range of centimeters to millimeters. Visible light is electromagnetic waves with much shorter wave lengths, ranging from 0.0004 millimeters for purple to 0.00055 millimeters for green to 0.0007 millimeters for red. Wavelengths between where visibility ends (0.0008 millimeters) and the microwave region begins (0.1 millimeters) are called infrared.

Earth's atmosphere contains molecules that absorb infrared radiation. They do not absorb the visible radiation coming in from the sun, so the Earth gets its full share of that. But a fraction of the infrared emitted by the Earth is absorbed by these molecules which then reemit it, frequently back to the Earth. That is what provides the extra heating. This is also the process that warms the plants in a greenhouse — the glass roof does not absorb the visible light coming in from the sun, but the infrared radiation emitted from the plants is absorbed by the glass and much of it is radiated back to the plants. That is how the process got its name — greenhouse effect. It is also the cause of automobiles getting hot when parked in the sun; the incoming visible radiation passes through the glass windows, while the infrared emitted from the car's interior is absorbed by the glass and much of it is emitted back into the interior.

Molecules in the atmosphere that absorb infrared and thereby increase the Earth's temperature are called greenhouse gases. Carbon dioxide is an efficient greenhouse gas. The atmosphere of Venus contains vast quantities of carbon dioxide, elevating its temperature by 500 degrees over what it would be without an atmosphere leading to nonconductive atmosphere to land on Venus.

Burning of coal, oil, and gas produces carbon dioxide, which adds to the supply already in the atmosphere, increasing the greenhouse effect and thereby increasing the temperature of the Earth. Prior to the industrial age, the concentration of carbon dioxide in the atmosphere was less than 280 ppm (parts per million). By 1958 the carbon dioxide concentration had risen to 315 ppm, and by 1986 it was 350 ppm. The average temperature of the Earth has been about 1 degree warmer in the 20th century than in the 19th century, which is close to what is expected from this carbon dioxide increase. As the rate of burning coal, oil, and gas escalates, so too does the rate of increase of carbon dioxide in the atmosphere.

Predicting the increase of temperature expected from this is very complicated, but since it is so important, a great deal of effort has gone into deriving estimates. Results are usually discussed in terms of doubling the concentration of carbon dioxide, from 350 ppm to 700 ppm. If current trends continue, this will occur during the next century, perhaps as early as 2030. The direct effect of doubling the carbon dioxide in the atmosphere would be to raise the Earth's average temperature by 2.2°F. Two side effects will accentuate this temperature rise. One is that the increased temperature causes more water to evaporate from the oceans, which adds to the number of water molecules in the atmosphere; water vapor is also a greenhouse gas. The other is that there would be less ice and snow; these reflect away the visible light from the sun that would otherwise be absorbed by the Earth's surface.

There are many other factors of lesser importance that must be considered. Some of these factors tend to reduce the warming effect:

- Clouds, generated by the increased evaporation of water, intercept some of the radiation coming in from the sun and emit part of it back into outer space.
- Volcanoes inject lots of dust into the atmosphere; this dust reflects sunlight away from the Earth.

- Plankton, tiny marine organisms whose growth is accelerated by carbon dioxide and higher temperatures, absorb carbon dioxide, thereby taking it out of circulation.
- Oceans absorb both carbon dioxide and heat.
- Sulfur dioxide, a pollutant we will be discussing soon, tends to cool the Earth, and in our efforts to eliminate pollution we are reducing this cooling effect.
- Thawing of permafrost, soil that has been frozen for thousands of years, releases methane (natural gas), which is a greenhouse gas.
- Bacteria in soil convert dead organic matter into carbon dioxide more rapidly as temperatures rise, thus increasing the amount of carbon dioxide in the atmosphere.

When all of these factors are taken into account as accurately as possible with our present knowledge, the best estimates are that doubling the carbon dioxide in the atmosphere will increase the average temperature by about 7°F.

The importance of this greenhouse effect has become a public issue because of the recent abnormally hot summers accompanied by droughts that have severely reduced our agricultural output.

4.4.4 Consequences of The Greenhouse Effect

Green house effects brings out damage to (i) Agriculture (ii) Forest (iii) Rise in sea level (iv) Wild Animal

Agriculture

Agriculture is especially sensitive to climate. For example, the hot, dry summer of 1988 reduced corn yields in the Midwest by 40%. But with long-term planning, crops can be changed to compensate for climate change. Moreover, increased levels of carbon dioxide would have beneficial effects on agriculture, since carbon dioxide in the air is the principal source of material from which plants produce food.

Live stock problems will increase. Heat stress will reduce breeding. Some of the live stock diseases that now plague the South will shift northward, and new tropical diseases will invade the South. Problems with agricultural pests will multiply. More pests will survive over the warmer winters, and they will breed more generations over the longer summers.

Forests

Forests will undergo some hard times. Each type of tree requires a specific climate. Thus, the growing area for each species will shift northward by 100 to 600 miles. This sounds innocuous, but adjustment periods will be difficult, with lots of die-off in the South and slow build-up in the North. Forests are constantly under stress from insects, diseases, competition with other plants, fires, wind, and the like. The added stress of changing climate is certain to cause trouble.

Rise in Sea Level

If all snow and ice were to melt, sea level would rise by 270 feet, enough to inundate nearly all of these cities and vast other areas of the nation. If present trends continue, there will be a 20-foot rise in 200-500 years. A reasonable estimate for the middle of the next century is 1.5-3 feet. A 3-foot rise would flood major areas of coast.

Inland penetration of salt water would cause lots of difficulties for aquatic life and Contamination of groundwater with salt would be a widespread problem.

Rising ocean levels do not necessarily mean rising levels in rivers and streams. For example, the levels of the Great Lakes are predicted to fall 2-5 feet, due to the greenhouse effect reducing rainfall and increasing evaporation. This will cause problems with shipping and with water supplies for cities and towns.

Wild Animal

Wild animals and plants must adapt to climate changes, and there are many potential difficulties. In some situations, animals can simply move, but not always. The grizzly bears in Yellowstone park would have no place to go and probably would die out. Other casualties of the greenhouse effect will probably be panthers, bald eagles, and spotted owls.

Insect plagues can be expected to cause lots of problems for trees, as will increased floods and droughts. Forest fires will occur more frequently.

Case Study

One would think that Canada would benefit from warmer climates, but there are many complications. With the oceans rising and the Great Lakes water levels falling, the St. Lawrence seaway will be in trouble. Southern Ontario, which has the most productive farmland in Canada, may suffer from drought as storm tracks and the rain they bring move north, and warmer temperatures cause increased evaporation. The western wheat belt will also be threatened by drought. But aside from these local problems, in general the greenhouse effect causes agriculture to move northward, and Canada can accommodate a lot of northward movement. The tree line moves north by about 35 miles for each degree Fahrenheit of global temperature rise.

As long as we burn fossil fuels, the Earth's climate will continue to get warmer. The only solution is to strongly reduce our burning of coal, oil, and gas and switching over to other non-polluting energy sources and renewable energy sources.

4.4.5 Acid Rain

Coal and oil contain small amounts of sulfur, typically 0.5% to 3% by weight. In the combustion process, sulfur combines with oxygen in the air to produce sulfur dioxide, which is the most important contributor to acid rain. Air consists of a mixture of oxygen (20%) and nitrogen (79%), and at very high temperatures molecules of these can combine to produce nitrogen oxides, the other important cause of acid rain. Sulfur dioxide and nitrogen oxides undergo chemical reactions in the atmosphere to become sulfuric acid and nitric acid, respectively, dissolved in water droplets that eventually may fall to the ground as rain. This rain is therefore acidic.

Naturally there is a substantial amount of carbon dioxide in the atmosphere. This dissolves in water droplets to form carbonic acid, familiar to us as carbonated water or soda water. As a result, "natural" rain is somewhat acidic, with a pH of about 5.6. Other natural factors, such as volcanic activity, also contribute, causing wide variations in pH. Rainfall pH as low as 4.0 has been observed even in places remote from the effects of fuel burning, like Antarctica and the Indian Ocean.

There is evidence that rain is made appreciably more acidic by the sulfuric and nitric acid from fossil fuel burning especially in the area of Gujarat and Maharastra.

Effects of Acid Rain

After the rain falls, it percolates through the ground, dissolving materials out of the soil. This alters its pH and introduces other materials into the water. If the soil is alkaline, the water's acidity will be neutralized, but if it is acid, the acidity of the water may increase. This water is used by plants and trees for their sustenance, and eventually flows into rivers and lakes. There have been various reports indicating that streams and lakes have been getting more acidic in recent years, although the effects seem to be highly variable and not closely correlated with releases of sulfur dioxide and nitrogen oxides. Among other thins, acid rain most frequently discussed are makes lakes unlivable for fish and other aquatic life and destroys forests.

Case Study

One of the problems is that it is difficult to be certain about any specific area. Of the 50,000 lakes in the northeastern United States, 220, mostly in the Adirondacks mountains of New York, have no fish because the water is too acidic. Since tourism is very important in the Adirondacks region, this has had serious economic consequences that may be ascribed to acid rain, and the residents are very upset about it. On the other hand, there is no evidence that there ever were fish in those lakes. The surrounding soil is naturally highly acidic, which is surely at least partially responsible for the lakes' acidity. Moreover, there is no indication that the acidity has been changing in recent years.

A study of one particular forest concluded that the trees were under stress from a variety of factors but acid rain was "the straw that broke the camel's back". It dissolved aluminum out of the soil – about 5% of all soil is aluminum – and this toxic element was picked up by the roots of trees. Not only did the toxicity of aluminum cause direct damage to the trees, but aluminum was picked up instead of calcium and magnesium, which are crucial to a tree's nutrition. The problem was compounded by the nitric acid in the rain acting as fertilizer to accelerate the trees' growth at a time when important nutrients were lacking. Because of this stress, the trees were succumbing to what would normally be non-lethal attacks by insects compounded by drought.

Trees are suffering blight in many parts of the world, and acid rain is suspected of contributing to the problem. A prime example are spruce trees in the Appalachian mountains, which are dying off from New England to North Carolina. In this situation, the trees are at a high altitude where they are engulfed in a mist a large fraction of the time. Acidity in the mist is believed to be an important contributor to the damage — acidity that has the same origin as acid rain, mainly emissions from coal-burning plants.

4.5 Ozone Layer Depletion

The distribution of ozone in the stratosphere is a function of altitude, latitude and season. It is determined by photochemical and transport processes. The ozone layer is located between 10 and 50 km above the Earth's surface and contains 90% of all stratospheric ozone. Under normal conditions, stratospheric ozone is formed by a photochemical reaction between oxygen molecules, oxygen atoms and solar radiation.

The ozone layer is essential to life on earth, as it absorbs harmful ultraviolet-B radiation from the sun. In recent years the thickness of this layer has been decreasing, leading in extreme cases to holes in the layer. Measurements carried out in the Antarctic have shown that at certain times, more than 95% of the ozone concentrations found at altitudes of between 15 and 20 km and more than 50% of total ozone are destroyed, with reductions being most pronounced during winter and in early spring. Natural phenomena, such as sun-spots and stratospheric winds, also decrease stratospheric ozone levels, but typically not by more than 1-2%.

The main cause of ozone layer depletion is the increased stratospheric concentration of chlorine from industrially produced CFCs , halons and selected solvents. Once in the stratosphere, every chlorine atom can destroy up to 100 000 ozone molecules. The amount of damage that an agent can do to the ozone layer is expressed relative to that of CFC-11 and is called the Ozone Depletion Potential (ODP), where the ODP of CFC-11 is 1.

Aircraft emissions of nitrogen oxides and water vapour add to this depletion effect by creating ice crystals that serve as a base for ozone destroying reactions.

The main potential consequences of this ozone depletion are:

- increase in UV-B radiation at ground level: a one percent loss of ozone leads to a two percent increase in UV radiation. Continuous exposure to UV radiation affects humans, animals and plants, and can lead to skin problems (ageing, cancer), depression of the immune system, and corneal cataracts (an eye disease that often leads to blindness). Increased UV radiation may also lead to a massive die-off of photoplancton (a CO_2 "sink") and therefore to increased global warming.
- disturbance of the thermal structure of the atmosphere, probably resulting in changes in atmospheric circulation;
- reduction of the ozone greenhouse effect: ozone is considered to be a greenhouse gas. A depleted ozone layer may partially dampen the greenhouse effect. Therefore efforts to tackle ozone depletion may result in increased global warming.

- changes in the tropospheric ozone and in the oxidising capacity of the troposphere.

International targets for the reduction of ozone depleting substances have resulted in the almost complete phasing out of CFCs, halons and carbon tetrachloride. Methyl chloroform and methyl bromide will be phased out by 2005 and HCFC by 2040.

4.6 Nuclear Accidents

The most dangerous thing in the use of Nuclear power or weapons is that when something goes wrong and an accident occurs, radiation is released into the environment which will lead to a lot of undesired effects.. So far many nuclear disasters have occurred and two of the most famous nuclear accidents are one at the Three Mile Island reactor 2 in the United States and the Chernobyl reactor 4 in the former Soviet Union. The details of these two case studies of disasters are as follows

Case study-1: Three Mile Island

Back round

Three Mile Island Nuclear Power Station is located on an island 10 miles from Harrisburg Pennsylvania. It is reported that there are two reactors at the plant, namely dubbed Unit 1 and Unit 2 and one of them is inoperable. Disaster occurred from the Unit 2 due to a partial reactor meltdown on March 28, 1979. A partial nuclear meltdown is when the uranium fuel rods start to liquefy, but they do not fall through the reactor floor and breach the containment systems. The accident which occurred at Unit 2 has been reported to be the worst nuclear disaster in US history and was due to mainly human error.

Causes of accident

The problem for the accident at Three Mile Island started at about four in the morning due to the failure of one of the control valves that control coolant flow into the reactor. So the core temperature started rising with the reduction in cooling water rate to reactor core. In the mean time the automatic computerized systems engaged, and the reactor was automatically scrammed. This led to stopping of chain reaction and hence slow down in the rise of reactor core temperature. However there is a residua heat in the reactor due decaying fission products.

Also there was a continuous removal of cooling water from reactor core with no supply of cooling water due to failure control valve leading to again rising in core temperature even emergency core cooling system was also put in to service automatically and this has not provided sufficient coolant with the wrong onion of the reactor operator that there is a enough coolant in the reactor.

In addition a valve at the top of the core automatically opened to vent some of the steam in the core providing little relief to rise in core temperature and did not close properly. This was not known to operator as valve indicator in the control was covered with maintenance tag attached to a nearby switch leading to wrong decision of situation is under control. After continuous venting of steam there was increase in rise core temperature leading to automatic switching on emergency core cooling. Even at this stage also there was not sufficient supply cooling water with the wrong opinion of the operator that the situation is under control.

With the continuous opening of vent valve led loss of coolant with the rise in core temperature. This temperature rise led to the collapse of fuel rod. At this stage the operator came to knew that something was wrong without knowing the reasons for it. Actually it took nearly two hours to know that valve releasing steam at the top of reactor did not close properly. The continuous removal of coolant led reactor meltdown. at 6.00 AM.

Explosion

During the day time of the same day build-up of hydrogen gas in the reactor led to an explosion in the after noon with no damage. After two days reactor core was not brought under by the operator. A group of experts were put into services & they pointed out that the hydrogen gas build-up was the reason for the explosion on the first day of the accident, or this build up could have displaced the remaining coolant in the reactor, causing a complete nuclear reactor meltdown. The reason for hydrogen build-up was not known even hydrogen recombiner was used to remove some of the hydrogen, but it was not very effective. Two weeks later the reactor was brought to a cold shutdown

No causality or injury has been reported as a result of the accident. However, some radioactive gas and water were vented to the environment around the reactor and also radioactive water was released into the Susquehanna River, which is a source of drinking water for nearby communities. Certainly there would have been some effects of these radioactive releases on people living near the power plant.

Case study-2: Chernobyl

Back round

About 80 miles (130 km) north of Kiev, in what is now the Ukraine, Chernobyl nuclear power plant is located Ukraine which is a part of Europe and one of the state ofF former USSR. At this plant the worst reactor disaster occurred on April 26, 1986. It is reported that some experiments was conduced in the actual reactor by suspending normal operations.without following normal safety guidelines that led to an accident.

History of accident

Early in the day, before the test, the power output of the reactor was dropped in preparation for the upcoming test. Unexpectedly, the reactor's power output dropped way too much, almost to zero. To make up this drop, some control rods were removed to bring the power back up. The reactor's power output raised up, and all appeared to be normal.

After the preparatory work, test trials were started and two pumps were switched on in the cooling system to increased water flow out of the reactor for faster removal of heat. The water level in a component of the reactor namely the steam separator has also been decreased

Due to combined action of many control rods removal and the reduction in water level, the amount of heat being taken from the core by the coolant had been reduced leading to dead hot condition and relatively low pressure in the core. Because of the heat and the low pressure, coolant inside the core began to boil to form steam. The actual test began with the closing of the turbine feed valves even there was a normal amount of steam in the reactor core lead to an surplus supply of steam and more reactor power output. The increase in power output lately realized

The temperature and pressure inside the reactor had already risen dramatically, and the fuel rods had begun to shatter leading to two explosions as a result of liquid uranium reacting with steam and fuel vapor expansion due to intense heat. The reactor containment was broken with entering of outside air in to the reactor. This air reacted with graphite moderator leading to the formation of CO. Carbon monoxide is flammable and soon caught fire resulting in extremely radioactive smoke into the area surrounding the reactor. Also, the explosion ejected a portion of the reactor fuel into the surrounding atmosphere and countryside. This fuel contained both fission products and transuranic wastes.

Remedial Measures taken

Liquid nitrogen was pumped into the reactor core to bring down the temperature. A large quantity of neutron-absorbing materials were dumped into the exposed core to prevent the reactor going critical conditions. Sand and other fire-fighting materials were also dumped into the core to stop the graphite fire. After the fires were completely brought under control, a big concrete seal was made around the reactor to prevent the release of any dangerous radiation

Effects of this disaster

The World Health Organization (WHO) reported that the radiation release from the Chernobyl accident was 200 times that of the combined effects of Hiroshima and Nagasaki nuclear bombs. Nearly 30 workers who were trying to put out the graphite fire and were exposed to radiation poisoning have lost their lives during the accident or within a few months after the accident. The radiation released has also had long-term effects on the cancer incidence rate of the surrounding population. According to the Ukrainian Radiological Institute over 2500 deaths resulted from the Chernobyl incident. The WHO has found a significant increase in cancer in the surrounding area. For example, in 1986 (the year of the accident), 2 cases of childhood thyroid cancer occurred in the Gomel administrative district of the Ukraine (this is the region around the plant). In 1993 there were 42 cases, which is 21 times the rate in 1986. The rate of thyroid cancer is particularly high after the Chernobyl accident because much of the radiation was emitted in the form of Iodine-131. Usually this Iodine-131 get collected in thyroid gland of particularly young children giving rise to thyroid cancer.

Major reason for the accident

The major cause of the accident is operator error with improper training. Also, normal safety rules were not being followed when the test running was carried out. For example, regulations required that at least 15 control rods always remain in the reactor. When the explosion occurred, less than 10 were present. This was one of major of the direct causes of the accident. It is also reported that the improper design of reactor led a possibility of abrupt and massive power surges.

4.7 Waste Reclamation

Approximately 20-25 % of the geographical area of India are now under the wasteland. Growing demands for fuel, fodder, wood and food has extensively depleted or eliminated protective plant cover and expressed surface soils to processes of degradation, resulting in partial to complete loss of soil and productivity. As a result the production of vegetation for food and other uses has extended to areas under great ecological stress and with less favorable environment. Mycorrhizal fungi are likely to be mot beneficial in diverse (Wasteland) ecosystems where the proportion of plants able to form Mycorrizhas is high and nutrient deficiencies are an important contribution to the rehabilitation of wasteland. through the following mechanisms:

Another outlook on reclamation is to use flyash as a special kind of soil. The ash characteristics particularly its heavy metal composition warrants the selection of plant species best suited for reclamation and meticulous screening of resistant Mycorrhizae.

Plantation of Bio-Diesel Produced from Plantations on Eroded Soils

The concept of substituting bio-diesel produced from plantations on eroded soils for conventional diesel fuel has gained widespread attention in India. In recent times, the

Indian central Government as well as some state governments have expressed their support for bringing marginal lands, which cannot be used for food production, under cultivation for this purpose. *Jatropha curcas* is a well established plant in India. It produces oil-rich seeds, is known to thrive on eroded lands, and to require only limited amounts of water, nutrients and capital inputs. This plant offers the option both to cultivate wastelands and to produce vegetable oil suitable for conversion to bio-diesel. More versatile than hydrogen and new propulsion systems such as fuel cell technology, bio-diesel can be used in today¿s vehicle fleets worldwide and may also offer a viable path to sustainable transportation, i.e., lower greenhouse gas emissions and enhanced mobility, even in remote areas. Mitigation of global warming and the creation of new regional employment opportunities can be important cornerstones of any forward looking transportation system for emerging economies.

Reclamation of a fly ash dump

Case studies

Two major fly ash reclamation projects have been successfully carried out in Badarpur (Delhi) and Korba (Chhattishgarh).

In Badarpur (Delhi), the reclamation of a fly ash dump (3900 square m) was successfully for the first time by introducing mycorrhizal technology at the Badarpur Thermal Power site. A variety of commercially viable plants, including tree species [Poplar (*Populus deltoides*), Sheesham (*Dalbergia sissoo*), *Eucalyptus*, and Meethi neem (*Melia azadirach*)] and floriculture and aromatic species (marigold, carnation, sunflower, lemongrass, tuberose, gladioli, and lily), could be able to grow using mycorrhizal technology.

It is reported that at the Korba Thermal Power site an approximately 8.44-acre of fly ash dump area has been reclaimed. The tree species used for the reclamation purpose have their own economic and commercial value, resulting in a long-term source of revenue generation. The tree species grown in the site is known for their high timber value. The species are (*Shorea robusta, Tectona grandis,* and *Dalbergia sissoo*), fodder (*Albizzia procera, Casuarina equisetifolia,* and *Melia azadriach*), tanin source (*Acacia nilotica*), paper and pulp industry (*Dendrocalamus strictus, Populus euphratica, Eucalyptus tereticornis*), plywood (*Bombex ceiba, Populus euphratica, Eucalyptus tereticornis*), and making musical instruments (*Gmelina arborea*). Commercially and medicinally important herbs like *Mentha arvensis* and *Vetiver zizanoides* have also been successfully grown in this fly ash dump and a lot of commercial products like oils are got from this area. It is also reported that ornamental plants like *Polianthes tuberosa, Helianthus* and *Tagetes erecta,* nitrogen fixer like *Sesbania aculeata* and for soil aggregation and Aloe gel (*Agave sisalana*) planted in this area are providing healthy atmosphere this area.

It is reported that FYM (Farm Yard Manure) was used in this project and this is sufficient for reclamation along with inter cropping and rotation of green manuring using appropriate mycorrhiza in both the sites. The advantages of this technology are (i) A significant restoration in soil quality with respect to percentage nitrogen, available phosphorus, available potassium, organic carbon, dehydrogenase activity and total microbial population has been recorded after two years of plantation activity (ii) Porosity and water holding capacity increased significantly over zero time (iii) the heavy metal toxicity in fly ash also decreased after two years of plantations (iv) The heavy metal content in the leachates also decreased drastically (v) the height of various tree species at Korba showed a tremendous improvement during the growth period.

4.8 Consumerism and Waste

Consumerism is economically manifested in the chronic purchasing of new goods and services, with little attention to their true need, durability, product origin or the environmental consequences of manufacture and disposal. Consumerism is driven by huge sums spent on advertising designed to create both a desire to follow trends, and the resultant personal self-reward system based on acquisition. Materialism is one of the end results of consumerism.

Environmental Concepts of Consumerism

Today new concepts are coming in to consumerism. More and more people are questioning a product's effect on the environment. Questions like;

Is my product ozone friendly?

How much waste does my product create?

Is its packaging recyclable?

Its my product toxic to its user or the environment?

How much energy is used in the making of the product?

The above are important to the customer due to general awareness of the Earth's rapidly increasing number of landfills, atmospheric pollution, water pollution, and the hole in the ozone layer.

The government and local bodies have taken action steps to recycle many products bought by the people. In many places, people have to pay to dispose of their garbage. Also, when you recycle bottles and other plastic products, you receive a small amount of money. This motivates people to take care of their environment. Many cities and towns are now encouraging people to practice the *4 R's:*

Reduce

Re-use

Renew

Recycle

Energy is not only used in the making of a product, but also in the use of certain products (electrical appliances, cars lawn mowers). The energy efficiency of a product is very important. To protect the environment, consumers must try to buy energy efficient products. For example, when purchasing a car, consumers should consider the mileage and emission control of the vehicle. The following are some examples of some products that are energy efficient, and some that are not ;

Energy Efficient	Non Energy Efficient
Floresence Lighting 17- 20%	Incandesance Lighting 5-7%
Diesel motor 40-45%	Gas motor 25-30%
Electric oven 95-99%	Batterie 85-90%

4.9 Forest (Conservation) Act, 1980 with Amendments Made In 1988

An Act to provide for the conservation of forests and for matters connected therewith or ancillary or incidental thereto.

Be it enacted by Parliament in the Thirty-first Year of the Republic of India as follows:-

1. Short Title, Extent and Commencement

(1) This Act may be called the Forest (Conservation) Act, 1980.

(2) It extends to the whole of India except the State of Jammu and Kashmir.

(3) It shall be deemed to have come into force on the 25th day of October, 1980.

2. Restriction on the Dereservation of Forests or Use of Forest Land for non-forest Purpose

Notwithstanding anything contained in any other law for the time being in force in a State, no State Government or other authority shall make, except with the prior approval of the Central Government, any order directing-

(i) that any reserved forest (within the meaning of the expression "reserved forest" in any law for the time being in force in that State) or any portion thereof, shall cease to be reserved;

(ii) that any forest land or any portion thereof may be used for any non-forest purpose;

(iii) that any forest land or any portion thereof may be assigned by way of lease or otherwise to any private person or to any authority, corporation, agency or any other organisation not owned, managed or controlled by Government;

(iv) that any forest land or any portion thereof may be cleared of trees which have grown naturally in that land or portion, for the purpose of using it for reafforestation.

Explanation : For the porpose of this section, "non-forest purpose" means the breaking up or clearing of any forest land or portion thereof for-

(a) the cultivation of tea, coffee, spices, rubber, palms, oil-bearing plants, horticultural crops or medicinal plants;

(b) any purpose other than reafforestation;

but does not include any work relating or ancillary to conservation, development and management of forests and wildlife, namely, the establishment of check-posts, fire lines, wireless communications and construction of fencing, bridges and culverts, dams, waterholes, trench marks, boundary marks, pipelines or other like purposes.

3. Constitution of Advisory Committee

The Central Government may constitute a Committee consisting of such number of persons as h may deem fit to advise that Government with regard to-

(i) the grant of approval. under Section 2; and

(ii) any other matter connected with the conservation of forests which may be referred to h by the Central Government.

3A. Penalty for contravention of the provisions of the Act.

Whoever contravenes or abets the contravention of any of the provisions of Section 2, shall be punishable with simple imprisonment for a period which may extend to fifteen days.

3B. Offences by the Authorities and Government Departments

(1) Where any offence under this Act has been committed-

(a) by any department of Government, the head of the department; or

(b) by any authority, every person who, at the time the offence was committed, was directly in charge of, and was responsible to, the authority for the conduct of the business of the authority as well as the authority;

shall be deemed to be guilty of the offence and shall be liable to be proceeded against and punished accordingly:

Provided that nothing contained in this sub-section shall render the head of the department or any person referred to in clause (b), liable to any punishment if he proves that the offence was committed without his knowledge or that he exercised all due diligence to prevent the commission of such offence.

(2) Notwithstanding anything contained in sub-section (1), where an offence punishable under the Act has been committed by a department of Government or any authority referred to in clause (b) of sub-section (1) and it is proved that the offence has been committed with the consent or connivance of; or is attributable to any neglect on the part of any officer, other than the head of the department, or in the case of an authority, any person other than the persons referred to in clause (b) of sub-section (1), such officer or persons shall also be deemed to be guilty of that offence and shall be liable to be proceeded against and punished accordingly.

4. Power to Make Rules

(1) The Central Government may, by notification in the Official Gazette, makes rules for carrying out the provisions of this Act.

(2) Every rule made under this Act shall be laid, as soon as may be after it is made, before each House of Parliament, while it is in session, for a total period of thirty days which may be comprised in one session or in two or more successive sessions, and if, before the expiry of the session immediately following the session or the successive sessions aforesaid, both Houses agree in making any modification in the rule or both Houses agree that the rule should not be made, the rule shall thereafter have effect only in such modified form or

be of no effect, as the case may be; so, however, that any such modification or annulment shall be without prejudice to the validity of anything previously done under that rule.

5. Repeal and Saving

(1) The Forest (Conservation) Ordinance, 1980 is hereby replaced.

(2) Notwithstanding such repeal, anything done or any action taken under the provisions of the said Ordinance shall be deemed to have been done or taken under the corresponding provisions of this Act.

4.10 Wildlife Protection Act

Government of India enacted a comprehensive legislation "**Wild Life (Protection) Act, 1972**" with the objective of effectively controlling poaching and illegal trade in wildlife and its derivatives. This has been amended recently (January, 2003) and punishment and penalty for offences under the Act have been made more stringent.

(i) Offences Pertaining to Hunting of Endangered Species and Altering of Boundaries of Protected Areas

For offences relating to wild animals (or their parts and products) included in schedule-I or part II of Schedule- II and those relating to hunting or altering the boundaries of a sanctuary or national park the punishment and penalty have been enhanced, the minimum imprisonment prescribed is three years which may extend to seven years, with a minimum fine of Rs. 10,000/-. For a subsequent offence of this nature, the term of imprisonment shall not be less than three years but may extend to seven years with a minimum fine of Rs. 25,000. Also a new section (51 - A) has been inserted in the Act, making certain conditions applicable while granting bail:

When any person accused of the commission of any offence relating to Schedule I or Part II of Schedule II or offences relating to hunting inside the boundaries of National Park or Wildlife Sanctuary or altering the boundaries of such parks and sanctuaries, is arrested under the provisions of the Act, then not withstanding anything contained in the Code of Criminal Procedure, 1973, no such person who had been previously convicted of an offence under this Act shall be released on bail unless -

(a) The Public Prosecutor has been given an opportunity of opposing the release on bail; and

(b) Where the Public Prosecutor opposes the application, the Court is satisfied that there are reasonable grounds for believing that he is not guilty of such offences and that he is not likely to commit any offence while on bail".

In order to improve the intelligence gathering in wildlife crime, the existing provision for rewarding the informers has been increased from 20% of the fine and composition money respectively to 50% in each case. In addition to this, a reward upto Rs. 10,000/- is also proposed to be given to the informants and others who provide assistance in detection of crime and apprehension of the offender.

At present, persons having ownership certificate in respect of Schedule I and Part II animals, can sell or gift such articles. This has been amended with a view to curb illegal trade, and thus no person can now acquire Schedule I or Part II of Schedule II animals, articles or trophies except by way of inheritance (except live elephants).

Stringent measures have also been proposed to forfeit the properties of hardcore criminals who have already been convicted in the past for heinous wildlife crimes. These provisions are similar to the provisions of 'Narcotic Drugs and Psychotropic Substances Act, 1985'. Provisions have also been made empowering officials to evict encroachments from Protected Areas.

(ii) Offences not Pertaining to Hunting of Endangered Species

Offences related to trade and commerce in trophies, animals articles etc. derived from certain animals (except chapter V A and section 38J) attracts a term of imprisonment upto three years and/or a fine upto Rs. 25,000/-.

4.11 The Air (Prevention and Control of Pollution) Act, 1981

An Act to provide for the prevention, control and abatement of air pollution, for the establishment, with a view to carrying out the aforesaid purposes, of Boards, for conferring on and assigning to such Boards powers and functions relating thereto and for matters connected therewith.

WHEREAS decisions were taken at the United Nations Conference on the Hum an Environment held in Stockholm in June, 1972, in which India participated, to take appropriate steps for the preservation of the natural resources of the earth which, among other things, include the preservation of the quality of air and control of air pollution;

AND WHEREAS it is considered necessary to implement the decisions aforesaid in so far as they relate to the preservation of the quality of air and control of air pollution;

This act consists of seven chapters and explained as follows

CHAPTER I Deals with preliminary like

1. Short title, extent and commencement.
2. Definitions relating to air pollution.

CHAPTER II Deals with central and state boards for the prevention and control of air pollution, it contains

3. Central Board for the Prevention and Control of Air Pollution.
4. State Boards for the Prevention and Control of Water Pollution to be, State Boards for the Prevention and Control of Air Pollution.
5. Constitution of State Boards.
6. Central Board to exercise the powers and perform die functions of a State Board in the Union territories.
7. Terms and conditions of service of members.
8. Disqualifications.
9. Vacation of seats by members.
10. Meetings-of Board.
11. Constitution -of committees.
12. Temporary association of persons with Board for particular purposes.
13. Vacancy in Board not to invalidate acts or proceedings.
14. Member-secretary and officers and other employees of State Boards.
15. Delegation of powers

CHAPTER III deals with the powers and functions of boards, it contains

16. Functions of Central Board.
17. Functions of State Boards.
18. Power to give directions.

CHAPTER IV deal with prevention and control of air pollution , it contains

19. Power to declare air pollution control areas,
20. Power to give instructions for ensuring standards for emission from automobiles.
21. Restrictions on use of certain industrial plants.
22. Persons carrying on industry, etc., and to allow emission of air pollutants in excess of the standard laid down by State Board.

22A. Power of Board to make application to court for restraining person from causing air pollution.

23. Furnishing, of information to State Board and other agencies in certain cases.
24. Power of entry and inspection.
25. Power to obtain information.
26. Power to take samples of air or emission and procedure to be followed in connection therewith.
27. Reports of the result of analysis on samples taken under section 26.
28. State Air Laboratory.
29. Analysis.
30. Reports of analysis.
31. Appeals,

CHAPTER V deals with fund, accounts and audit, it contains

32. Contribution by Central Government.
33. Fund of Board.
33A. Borrowing powers of Board.
34. Buduct.
35. Annual report.
36. Accounts and audit.

CHAPTER VI deals with penalties and procedure, it conatins

37. Failure to comply with the provisions of section 21 or section 22 or with the directions issued under section 31A.
38. Penalties for certain acts.
39. Penalty for contravention of provisions of the Act.
40. Offences by companies.
41. Offences by Government Departments.
42. Protection of action taken in good faith
43. Cognizance of offences
44. Members, officers and employees of Board to be public servants.
45. Reports and returns.
46. Bar of jurisdiction.

CHAPTER VII deals with miscellaneous information relation to air pollution, it contains

47. Power of Central Government to supersede State Board,
48. Special provision in the case of supersession of the Central Board or the State Boards constituted under

the Water (Prevention and Control of Pollution) Act, 1974.

49. Dissolution of State Boards constituted under the Act
50. [Power to amend the Schedule.] Rep. by the Air (Prevention and Control of Pollution) Amendment Act, 1987 (47 of 1987), s. 22 (w.e.f. 1-41988).
51. Maintenance of register.
52. Effect of other laws.
53. Power of Central Government to make rules.
54. Power of State Government to make rules.

Details of this is act in given in Annexure-I

4.12 Water (Prevention and Control of Pollution) Act, 1974

This is an Act to provide for the prevention and control of water pollution and the maintaining or restoring of wholesomeness of water for the establishment, with a view to carrying out the purposes aforesaid, of Boards for the prevention and control of water pollution, for conferring on and assigning to such Boards powers and functions relating thereto and for matters connected therewith

WHEREAS it is expedient to provide for the prevention and control of water pollution and the maintaining or restoring of wholesomeness of water, for the establishment, with a view to carrying out the purposes aforesaid, of Boards for the prevention and control of water pollution and for conferring on and assigning to such Boards powers and functions relating thereto;

AND WHEREAS Parliament has no power to make laws for the States with respect to any of the matters aforesaid except as provided in articles 249 and 250 of the Constitution;

AND WHEREAS in pursuance of clause (1) of article 252 of the Constitution resolutions have been passed by all the Houses of the Legislatures of the States of Assam, Bihar, Gujarat, Haryana, Himachal Pradesh, Jammu and Kashmir, Karnataka, Kerala, Madhya Pradesh, Rajasthan, Tripura and West Bengal to the effect that the matters aforesaid should be regulated in those States by Parliament by law;

It was enacted by Parliament in the Twenty-fifth Year of the Republic of India as follows: The act details contain eight chapters

CHAPTER I Deals with Preliminary information like

1. Short title, application and commencement
2. Definitions

CHAPTER II Deals with the central and state boards for prevention and control of water pollution, and contains information on

3. Constitution of Central Board
4. Constitution of State Boards
5. Terms and conditions of service of members
6. Disqualifications
7. Vacation of seats by members
8. Meetings of Boards
9. Constitution of committees
10. Temporary association of persons with Board for particular purposes
11. Vacancy in Board not to invalidate acts or proceedings
12. Member-secretary and officers and other employees of Board

CHAPTER III Deals with joint boards like state and central pollution control boards and giving information on

13. Constitution of Joint Boards
14. Composition of Joint Boards
15. Special provisions relating to giving of directions

CHAPTER IV Deals with powers and functions of boards and gives information on

16. functions of central board
17. functions of state board
18. power to give directions

CHAPTER V Deals with prevention and control of water pollution norms and provide informations like

19. Power of State Government to restrict the application of the Act to certain areas
20. Power to obtain information
21. Power to take samples of effluents and procedure to be followed in connection therewith
22. Reports of the result of analysis on samples taken under section 21
23. Power of entry and inspection
24. Prohibition on use of stream or well for disposal of polluting matter, etc.
25. Restrictions on new outlets and new discharges
26. Provision regarding existing discharge of sewage or trade effluent
27. Refusal or withdrawal of consent by State Board
28. Appeals

29. Revision
30. Power of State Board to carry out certain works
31. Furnishing of information to State Board and other agencies in certain cases
32. Emergency measures in case of pollution of stream or well
33. Power of Board to make application to courts for restraining apprehended pollution of water in streams or wells

CHAPTER VI Deals with funds, accounts and audit of pollution control board

34. Contributions by Central Government
35. Contributions by State Government
36. Fund of Central Board
37. Fund of State Board
38. Budget
39. Annual report
40. Accounts and audit

CHAPTER VII Deal with penalties and procedure and provide informations like

41. Failure to comply with directions under sub-section (2) or sub-section (3) of section 20
42. Penalty for certain acts
43. Penalty for contravention of provisions of section 24
44. Penalty for contravention of section 25 or section 26
45. Enhanced penalty after previous conviction
46. Publication of names of offenders

47. Offences by companies
48. Offences by government departments
49. Cognizance of offences
50. Members, officers and servants of Board to be public servants

CHAPTER VIII Gives miscellaneous informations of the following

51. Central Water Laboratory
52. State Water Laboratory
53. Analysts
54. Reports of analysts
55. Local authorities to assist
56. Compulsory acquisition of land for the State Board
57. Returns and reports
58. Bar of jurisdiction
59. Protection of action taken in good faith
60. Overriding effect
61. Power of Central Government to supersede the Central Board and Joint Boards
62. Power of State Government to supersede State Board
63. Power of Central Government to make rules
64. Power of State Government to make rules

Details of this is given in Annexure-II

4.13 The Environment (Protection) Act, 1986

The Environment Protection Act aims to ensure the protection and continuous improvement of the environment by controlling and preventing environmental pollution. The

Government prescribes guidelines and makes rules on the quality standards of air, water and soil for different areas and the maximum levels of emission, to which industries must adhere.

An Act to provide for the protection and improvement of environment and for matters connected there with:

This act was enacted by Parliament in the Thirty-seventh Year of the Republic of India.

This act contains four chapters.

Chapter I Deals with Applicability and basic definitions related Environment,

Chapter II Deals with

General Powers of the central government

Power of central government to take measures to protect and improve environment

Appointment of officers and their powers and functions

Power to give directions

Rules to regulate environmental pollution

Chapter III Contains

Prevention, control, and abatement of environmental pollution

Persons carrying on industry operation, etc., not to allow emission or discharge of environmental pollutants in excess of the standards

Persons handling hazardous substances to comply with procedural safeguards

Furnishing of information to authorities and agencies in

Powers of entry and inspection

Power to take sample and procedure to be followed in connection therewith

Environmental laboratories

Government analysts

Reports of government analysts

Penalty for contravention of the provisions of the act and the rules, orders and directions

Offences by companies

Offences by government departments

CHAPTER IV Deals with

Miscellaneous information

Protection of action taken in good faith

Cognizance of offences

Information, reports or returns

Members, officers and employees of the authority constituted under section 3 to be public servants

Bar of jurisdiction

Powers to delegate

Effect of other laws

Power to make rules

Rules made under this act to be laid before parliament

Details of this act is given in Annexure -III

4.14 Environmental Laws

In the Constitution of India it is clearly stated that it is the duty of the state to 'protect and improve the environment and to safeguard the forests and wildlife of the country'. It imposes a duty on every citizen 'to protect and improve the natural environment including forests, lakes, rivers, and wildlife'. Reference to the environment has also been made in the Directive Principles of State Policy as well as the Fundamental Rights. The Department of Environment was established in India in 1980 to ensure a healthy environment for the country. This later became the Ministry of Environment and Forests in 1985.

The constitutional provisions are backed by a number of laws - acts, rules, and notifications. The EPA (Environment Protection Act), 1986

This can be considered as:

General

Forest and wildlife

Water

Air

General

1986 - The Environment (Protection) Act authorizes the central government to protect and improve environmental quality, control and reduce pollution from all sources, and prohibit or restrict the setting and /or operation of any industrial facility on environmental grounds.

1986 - The Environment (Protection) Rules lay down procedures for setting standards of emission or discharge of environmental pollutants.

1989 - The objective of Hazardous Waste (Management and Handling) Rules is to control the generation, collection, treatment, import, storage, and handling of hazardous waste.

1989 - The Manufacture, Storage, and Import of Hazardous Rules define the terms used in this context, and sets up an authority to inspect, once a year, the industrial activity connected with hazardous chemicals and isolated storage facilities.

1989 - The Manufacture, Use, Import, Export, and Storage of hazardous Micro-organisms/ Genetically Engineered Organisms or Cells Rules were introduced with a view to protect the environment, nature, and health, in connection with the application of gene technology and microorganisms.

1991 - The Public Liability Insurance Act and Rules and Amendment, 1992 was drawn up to provide for public liability insurance for the purpose of providing immediate relief to the persons affected by accident while handling any hazardous substance.

1995 - The National Environmental Tribunal Act has been created to award compensation for damages to persons, property, and the environment arising from any activity involving hazardous substances.

1997 - The National Environment Appellate Authority Act has been created to hear appeals with respect to restrictions of areas in which classes of industries etc. are carried out or prescribed subject to certain safeguards under the EPA.

1998 - The Biomedical waste (Management and Handling) Rules is a legal binding on the health care institutions to streamline the process of proper handling of hospital waste such as segregation, disposal, collection, and treatment.

1999 - The Environment (Siting for Industrial Projects) Rules, 1999 lay down detailed provisions relating to areas to be avoided for siting of industries, precautionary measures to be taken for site selecting as also the aspects of environmental protection which should have been incorporated during the implementation of the industrial development projects.

2000 - The Municipal Solid Wastes (Management and Handling) Rules, 2000 apply to every municipal authority responsible for the collection, segregation, storage, transportation, processing, and disposal of municipal solid wastes.

2000 - The Ozone Depleting Substances (Regulation and Control) Rules have been laid down for the regulation of production and consumption of ozone depleting substances.

2001 - The Batteries (Management and Handling) Rules, 2001 rules shall apply to every manufacturer, importer, re-conditioner, assembler, dealer, auctioneer, consumer, and bulk consumer involved in the manufacture, processing, sale, purchase, and use of batteries or components so as to regulate and ensure the environmentally safe disposal of used batteries.

2002 - The Noise Pollution (Regulation and Control) (Amendment) Rules lay down such terms and conditions as are necessary to reduce noise pollution, permit use of loud speakers or public address systems during night hours (between 10:00 p.m. to 12:00 midnight) on or during any cultural or religious festive occasion.

2002 - The Biological Diversity Act is an act to provide for the conservation of biological diversity, sustainable use of its components, and fair and equitable sharing of the benefits arising out of the use of biological resources and knowledge associated with it.

Forest and Wildlife

1927 - The Indian Forest Act and Amendment, 1984, is one of the many surviving colonial statutes. It was enacted to 'consolidate the law related to forest, the transit of forest produce, and the duty leviable on timber and other forest produce'.

1972 - The Wildlife Protection Act, Rules 1973 and Amendment 1991 provides for the protection of birds and animals and for all matters that are connected to it whether it be their habitat or the waterhole or the forests that sustain them.

1980 - The Forest (Conservation) Act and Rules, 1981, provides for the protection of and the conservation of the forests.

Water

1882 - The Easement Act allows private rights to use a resource that is, groundwater, by viewing it as an attachment to the land. It also states that all surface water belongs to the state and is a state property.

1897 - The Indian Fisheries Act establishes two sets of penal offences whereby the government can sue any person who uses dynamite or other explosive substance in any way (whether coastal or inland) with intent to catch or destroy any fish or poisonous fish in order to kill.

1956 - The River Boards Act enables the states to enroll the central government in setting up an Advisory River Board to resolve issues in inter-state cooperation.

1970 - The Merchant Shipping Act aims to deal with waste arising from ships along the coastal areas within a specified radius.

1974 - The Water (Prevention and Control of Pollution) Act establishes an institutional structure for preventing and abating water pollution. It establishes standards for water quality and effluent. Polluting industries must seek permission to discharge waste into effluent bodies.

The CPCB (Central Pollution Control Board) was constituted under this act.

1977 - The Water (Prevention and Control of Pollution) Cess Act provides for the levy and collection of cess or fees on water consuming industries and local authorities.

1978 - The Water (Prevention and Control of Pollution) Cess Rules contains the standard definitions and indicate the kind of and location of meters that every consumer of water is required to affix.

1991 - The Coastal Regulation Zone Notification puts regulations on various activities, including construction, are regulated. It gives some protection to the backwaters and estuaries.

Air

1948 – The Factories Act and Amendment in 1987 was the first to express concern for the working environment of the workers. The amendment of 1987 has sharpened its environmental focus and expanded its application to hazardous processes.

1981 - The Air (Prevention and Control of Pollution) Act provides for the control and abatement of air pollution. It entrusts the power of enforcing this act to the CPCB .

1982 - The Air (Prevention and Control of Pollution) Rules defines the procedures of the meetings of the Boards and the powers entrusted to them.

1982 - The Atomic Energy Act deals with the radioactive waste.

1987 - The Air (Prevention and Control of Pollution) Amendment Act empowers the central and state pollution control boards to meet with grave emergencies of air pollution.

1988 - The Motor Vehicles Act states that all hazardous waste is to be properly packaged, labelled, and transported.

Test Your Understanding

1. Discuss various urban issue in the Modern life
2. What is Environmental ethics? Explain
3. What is water conservation? Explain various methods
4. Write short notes on

   (i) Global warming

   (ii) Acid rain

   (iii) Ozone layer depletion

   (iv) Green House effect

   (v) Consumerism and waste
5. Explain various features forest conservation Act
6. Discuss various features in Air Pollution prevention and control Act

   Explain the various features of water pollution prevention and control Act

CHAPTER 5

Human Population and The Environment

5.1 Population Growth

In 2000, the world had 6.1 billion human inhabitants. This figure may rise to more than 9 billion in the next 50 years. For the last 50 years, world population multiplied more rapidly than ever before, and more rapidly than it will ever grow in the future. Anthropologists believe the human species dates back at least 3 million years. For most of our history, these distant ancestors lived a precarious existence as hunters and gatherers. This way of life kept their total numbers small, probably less than 10 million. However, as agriculture was introduced, communities evolved that could support more people.

World population expanded to about 300 million by A.D. 1 and continued to grow at a moderate rate. But after the start of the Industrial Revolution in the 18th century, living standards rose and widespread famines and epidemics diminished in some regions. Population growth accelerated. The population climbed to about 760 million in 1750 and reached 1 billion around 1800.

In 1800, the vast majority of the world's population (86 percent) resided in Asia and Europe, with 65 percent in Asia alone. By 1900, Europe's share of world population had risen to 25 percent, fueled by the population increase that accompanied the Industrial Revolution. Some of this growth spilled over to the Americas, increasing their share of the world total. The 2000 growth rate of 1.4 percent, when applied to the world's 6.1 billion population, yields an annual increase of about 85 million people. Because of the large and increasing population size, the number of people added to the global population will remain high for several decades, even as growth rates continue to decline.

Between 2000 and 2030, nearly 100 percent of this annual growth will occur in the less developed countries in Africa, Asia, and Latin America, whose population growth rates are much higher than those in more developed countries.

5.2 Exponential Growth

As long ago as 1789, Thomas Malthus studied the nature of population growth in Europe. He claimed that population was increasing faster than food production, and he feared eventual global starvation. Of course he could not foresee how modern technology would expand food production, but his observations about how populations increase were important. Population grows geometrically (1, 2, 4, 8 ...), rather than arithmetically (1, 2, 3, 4 ...), leading to sharp increase in number.

Similarly, if a country's population begins with 1 million and grows at a steady 3 percent annually, it will add 30,000 persons the first year, almost 31,000 the second year, and 40,000 by the 10th year. At a 3 percent growth rate, its doubling time – or the number of years to double in size – is 23 years. (The doubling time for a population can be roughly determined by dividing the current growth rate into the number "69." Therefore, 69/3=23 years. Of course, if a population's growth rate does not remain at this rate, the projected doubling time would need to be recalculated.)

The 2000 growth rate of 1.4 percent, when applied to the world's 6.1 billion population, yields an annual increase of about 85 million people. Because of the large and increasing population size, the number of people added to the global population will remain high for several decades, even as growth rates continue to decline.

Between 2000 and 2030, nearly 100 percent of this annual growth will occur in the less developed countries in Africa, Asia, and Latin America, whose population growth rates are much higher than those in more developed countries. Growth rates of 1.9 percent and higher mean that populations would double in about 36 years, if these rates continue. Demographers do not believe they will. Projections of growth rates are lower than 1.9 percent because birth rates are declining and are expected to continue to do so. The populations in the less developed regions will most likely continue to command a larger proportion of the world total. While Asia's share of world population may continue to hover around 55 percent through the next century, Europe's portion has declined sharply and could drop even more during the 21st century. Africa and Latin America each would gain part of Europe's portion. By 2100, Africa is expected to capture the greatest share (see chart, "World population distribution by region, 1800–2050", above).

The more developed countries in Europe and North America, as well as Japan, Australia, and New Zealand, are growing by less than 1 percent annually. Population growth rates are negative in many European countries, including Russia (-0.6%), Estonia (-0.5%), Hungary (-0.4%), and Ukraine (-0.4%). If the growth rates in these countries continue to fall below zero, population size would slowly decline. As the chart "World population growth, 1750–2150" shows, population increase in more developed countries is already low and is expected to stabilize.

5.3 Population Explosions

World population growth accelerated after World War II, when the population of less developed countries began to increase dramatically. After millions of years of extremely slow growth, the human population indeed grew explosively, doubling again and again; a billion people were added between 1960 and 1975; another billion were added between 1975 and 1987. Throughout the 20th century each additional billion has been achieved in a shorter period of time. Human population entered the 20th century with 1.6 billion people and left the century with 6.1 billion. The growth of the last 200 years appears explosive on the historical timeline. The overall effects of this growth on living standards, resource use, and the environment will continue to change the world landscape long after.

5.4 Over Population

In the past, infant and childhood deaths and short life spans used to limit population growth. In today's world, thanks to improved nutrition, sanitation, and medical care, more babies survive their first few years of life. The combination of a continuing high birth rate and a low death rate is creating a rapid population increase in many countries in Asia, Latin America. Over-population is defined as the condition of having more people than can live on the earth in comfort, happiness and health and still leave the world a fit place for future generations. The greatest threat to the future comes from overpopulation.

It took the entire history of humankind for the population to reach 1 billion around 1810. Just 120 years later, this doubled to 2 billion people (1930); then 4 billion in 1975 within span of 45 years. The number of people in the world has increased from 4.4 billion people in 1980 to 5.8 billion today. It is estimated that the population could double again to nearly 11 billion in less than 40 years. This means that more people are now being added each day than at any other time in human history.

Looking ahead, world population is projected to exceed 6 billion before the year 2000. And according to a report by the United Nation Population fund, total population is likely to reach 10 billion by 2025 and grow to 14 billion by the end of the next century unless birth control use increases dramatically around the world within the next two decades.

Both death rates and birth rates have fallen, but death rates have fallen faster than birth rates. There are about 3 births for each death with 1.6 births for each death in more developed countries (MDCs) and 3.3 births for each death in less developed countries(LDCs). The world's population continues to grow by 1 billion people every dozen years.

On one hand, some politicians call for countries, especially MDCs to increase their population size to maintain their economic growth and military security. On the other hand, critics denote that one out of five people living here today is not properly supported and believe that the world is already limited in resources.

5.5 The Causes of Rapid Population Growth

Until recently, birth rates and death rates were about the same, keeping the population stable. People had many children, but a large number of them died before age five. During the Industrial Revolution, a period of history in Europe and North America where there were great advances in science and technology, the success in reducing death rates was attributable to several factors: (1) in-creases in food production and distribution, (2) improvement in public health (water and sanitation), and (3) medical technology (vaccines and antibiotics), along with gains in education and standards of living within many developing nations. In addition, because of the technology, people could produce more and different kinds of food. Gradually, over a period of time, these discoveries and inventions spread throughout the world, lowering death rates and improving the quality of life for most people.

5.6 Ways and Means to Solve Over Population

Population projections represent the playing out into the future of a set of assumptions about future fertility and mortality rates. More public education is needed to develop more awareness about population issues. Facts like the size or the growth rate of the human population should be in the head of every citizen. Schools should inform students about population issues in order for them to make projections about the future generations.

Ways and means can be developed to increase public understanding of how rapid population growth limits chances for meeting basic needs. The spirit of open communication and empowerment of individual women and men will be key to a successful solution to many population problems. Collective vision about health care, family planning and women's education at the community level build a basis for action. The creation of action plans help to meet challenges to find cooperative solutions. Free and equal access to health care, family planning and education are desirable in their own right and will also help reduce unwanted fertility.

Individual choice, human rights and collective responsibility are key to allowing families to plan the size and spacing of their children. It is essential to achieve a balance between population and the available resources. Teachers, parents, community workers and other stakeholders should extend the range of choices about available resources to individuals, especially women, and by equalizing opportunities between the genders from birth onwards. Teachers, parents, other educators, politicians and other concerned citizens can practice how to make good decisions in everyday life. Decisions about family size and resource will affect the future generations. Through community forums, specific issues about the population growth can be discussed and possible action plans can be developed. The investigation of world population will raise the level of awareness, so that we can learn to handle problems based on data.

5.7 Environment and Health

Human health is intimately connected to the surrounding environment - every day most of us breathe in air pollution, digest artificial chemicals with our food, and experience stressful noise levels, to name just three examples. It is usually very difficult to identify cause-and-effect relationships between, say, noise pollution and heart disease. Identifying these relationships, however, is often worth the effort – linking environmental pollution to human health helps to redefine priorities and unlock resources

On our planet, air is one of the most important natural resources on which all life depends. However, our atmosphere is also in the front line for receiving environmental pollution.

Noise is a serious issue - levels over 40 Ldn dB(A) affect our well-being, while there is evidence that levels over 60 Ldn dB(A) can affect our physical and psychological health.

While the last two decades have seen significant reductions in the noise produced by all sorts of vehicles, the rapid growth in transport – particularly air and road

Water: Clean fresh water is essential to life. Unfortunately, since the Industrial Revolution, most of Europe's rivers have been treated more like a convenient way of transporting waste to the sea, destroying the biodiversity of thousands of kilometers of waterways, harming human health, and polluting coastal waters in the process

The past decades have seen significant progress in treating the sewage and industrial wastes which are being pumped into Europe's river systems, resulting in lower levels of most pollutants and a measurable improvement in water quality. The agricultural sector, on the other hand, has not made as much progress. Nitrate levels in Europe's rivers are still as high as they were at the beginning of the last decade.

Not only the quality of water but also the quantity available for human use is of importance and more and more frequently, there are problems with water scarcity around large cities and in many parts of the world.

5.8 Family Welfare Programme

India launched the National Family Welfare Programme in 1951 with the objective of "reducing the birth rate to the extent necessary to stabilise the population at a level consistent with the requirement of the National economy.

"The Family Welfare Programme in India is recognized as a priority area, and is being implemented as a 100% centrally sponsored programme. As per Constitution of India, Family Planning is in the Concurrent list. The approach under the programme during the First and Second Five Year Plans was mainly "Clinical" under which facilities for provision of services were created. However, on the basis of data brought out by the 1961 census, clinical approach adopted in the first two plans was replaced by "Extension and Education Approach" which envisaged expansion of services facilities along with spread of message of small family norm.

iv Five Year Plan

In the IV Plan (1969-74), high priority was accorded to the programme and it was proposed to reduce birth rate from 35 per thousand to 32 per thousand by the end of plan. 16.5 million couples, constituting about 16.5% of the couples in the reproductive age group, were protected against conception by the end of IVth Plan.

v Five Year Plan

The objective of the V plan (1974-79) was to bring down the birth rate to 30 per thousand by the end of 1978-79. The programme was included as a priority sector programme during the V Plan with increasing integration of family planning services with those of Health, Maternal and Child Health (MCH) and nutrition, so that the programme became more readily acceptable. The years 1975-76 and 1976-77 recorded a phenomenal increase in performance of sterilization. However, in view of rigidity in enforcement of targets by field functionaries and an element of coercion in the implementation of the programme in 1976-77 in some areas, the programme received a set-back during 1977-78. As a result, the Government made it clear that there was no place for force or coercion or compulsion or for pressure of any sort under the programme and the programme had to be implemented as an integral part of "Family Welfare" relying solely on mass education and motivation. The name of the programme also was changed to Family Welfare from Family Planning. The change was not merely in nomenclature but essentially in the content of its objectives.

vi Five Year Plan

In the VI Plan (1980-85), certain long-term demographic goals of reaching net reproduction rate of unity were envisaged. The implications of this were to achieve the following by the year 2000 AD.

1. Reduction of average size of family from 4.4 children in 1975 to 2.3 children.
2. Reduction of birth rate to 21 from the level of 33 in 1978 and death rate from 14 to 9 and infant mortality rate from 127 to below 60.
3. Increasing the couple protection level from 22% to 60%.

Year-wise achievement during the VI Plan period of the four Family Planning Methods was as below: (Figures in 1000)

Year	Sterilisations	IUD Insertions	C.C	Oral Pills
1980-81	2053	628	3718	91
1981-82	2792	751	4439	120
1982-83	3983	1097	5765	183
1983-84	4532	2134	7661	729
1984-85	4085	2562	8505	1290

vii Five Year Plan

The Family Welfare Programme during VII five year plan (1985-90) was continued on a purely voluntary basis with emphasis on promoting spacing methods, securing maximum community participation and promoting maternal and child health care. In order to provide facilities/services nearer to the door steps of population, the following steps/initiatives were taken during the VII Plan period.

1. It was envisaged to have one sub-centre for every 5000 population in plain areas and for 3000 population in hilly and tribal areas. At the end of VII plan

i.e.31.3.1990, 1.30 lakhs sub-centres were established in the country.

2. The Post Partum programme was progressively extended to sub-district level hospitals. At the end of VII plan, 1012 sub-district level hospitals and 870 Health Posts were established in the country.
3. The Universal Immunization Programme started in 30 Districts in 1985-86 was extended to cover all the districts in the country by the end of the VII plan.
4. A project for improving Primary Health Care in urban slums in the cities of Bombay and Madras was taken up with assistance from World Bank.
5. Area Development Projects were implemented in selected districts of 15 major States with assistance from various donor Agencies.

The achievements of the Family Welfare Programme at the end of the VII plan were

(i) Reduction in crude birth rate from 41.7 (1951-61) to 30.2 (SRS:1990).

(ii) Reduction in total fertility rate from 5.97 (1950-55) to 3.8 (SRS:1990).

(iii) Reduction in infant mortality rate from 146 (1970-71) to 80 (SRS:1990).

(iv) Increase in Couple Protection Rate from 10.4% (1970-71) to 43.3% (31.3.1990).

(v) Setting up of a large network of service delivery infrastructure, which was virtually non-existent at the inception of the programme.

(vi) Over 118 million births were averted by the end of March, 1990.

viii Five Year Plan

To impart new dynamism to the Family Welfare Programme, several new initiatives were introduced and ongoing schemes were revamped in the Eighth Plan (1992-97). The broad features of these initiatives are as under:

World Bank assisted Area Projects which seek to upgrade infrastructure and development of trained manpower have been continued during the 8th Five Year Plan. Two new Area Projects namely India Population Project (IPP)-VIII and IX have been initiated during the 8th Plan. The IPP-VIII project aims at improving health & family welfare services in the urban slums in the cities of Delhi, Calcutta, Hyderabad and Bangalore. IPP-IX will operate in the States of Rajasthan, Assam and Karnataka.

An USAID assisted project named "Innovations in Family Planning Services" has been taken up in Uttar Pradesh with specific objective of reducing TFR from 5.4 to 4 and increasing CPR from 35% to 50% over the 10 years project period.

Recognising the fact that demographic and health profile of the country is not uniform, 90 districts which have CBR of over 39 per thousand (1991 census) were identified for differential programming. Enhanced allocation of financial resources, amounting to Rs.50 lakhs per year per district, was made for upgradation of health infrastructure in these districts from 1992-93 to 1995-96. This amount is being used for providing well equipped Operation Theatres, Labour Room, a six-bedded observation ward and residential quarters for paramedical workers in 5 PHCs of each district per year. All the block level PHCs of these 90 districts have been covered.

Realising that Government efforts alone in propagating and motivating the people for adaptation of small family norm would not be sufficient, greater stress has been laid on the involvement of NGOs to supplement and complement the Government efforts. Four new schemes for increasing the involvement of NGOs have been evolved by the Department of Family Welfare.

The Universal Immunisation Programme (UIP) was launched in 1985 to provide universal coverage of infants and pregnant women with immunisation against identified vaccine preventable diseases. From the year 1992-93, the UIP has been strengthened and expanded into the Child Survival and Safe Motherhood (CSSM) Project. It involves sustaining the high immunisation coverage level under UIP, and augmenting activities under Oral Rehydration Therapy, prophylaxis for control of blindness in children and control of acute respiratory infections. Under the Safe Motherhood component, training of traditional birth attendants, provision of aseptic delivery kits and strengthening of first referral units to deal with high risk and obstetric emergencies are being taken up.

The targets fixed for the 8th plan of a National level birth rate of 26 was achieved by all States except the States of Assam, Bihar, Haryana, Madhya Pradesh, Orissa, Rajasthan and Uttar Pradesh.

ix. Five Year Plan (1997-2002)

Reduction in the population growth rate has been recognised as one of the priority objectives during the Ninth Plan period.

The objectives during the Ninth Plan are:

i) to meet all the felt-needs for contraception

ii) to reduce the infant and maternal morbidity and mortality so that there is a reduction in the desired level of fertility.

The strategies during the Ninth Plan will be:

i) to assess the needs for reproductive and child health at PHC level and undertake area-specific micro planning.

ii) To provide need-based, demand-driven, high quality, integrated reproductive and child health care.

5.9 Human Rights

On December 10, 1948 the General Assembly of the United Nations adopted and proclaimed the Universal Declaration of Human Rights the full text of which appears in the following pages. Following this historic act the Assembly called upon all Member countries to publicize the text of the Declaration and "to cause it to be disseminated, displayed, read and expounded principally in schools and other educational institutions, without distinction based on the political status of countries or territories."

Preamble

Whereas recognition of the inherent dignity and of the equal and inalienable rights of all members of the human family is the foundation of freedom, justice and peace in the world,

Whereas disregard and contempt for human rights have resulted in barbarous acts which have outraged the conscience of mankind, and the advent of a world in which human beings shall enjoy freedom of speech and belief and freedom from fear and want has been proclaimed as the highest aspiration of the common people,

Whereas it is essential, if man is not to be compelled to have recourse, as a last resort, to rebellion against tyranny and oppression, that human rights should be protected by the rule of law,

Whereas it is essential to promote the development of friendly relations between nations,

Whereas the peoples of the United Nations have in the Charter reaffirmed their faith in fundamental human rights, in the dignity and worth of the human person and in the equal rights of men and women and have determined to promote social progress and better standards of life in larger freedom,

Whereas Member States have pledged themselves to achieve, in co-operation with the United Nations, the promotion of universal respect for and observance of human rights and fundamental freedoms,

Whereas a common understanding of these rights and freedoms is of the greatest importance for the full realization of this pledge,

Now, Therefore THE GENERAL ASSEMBLY proclaims THIS UNIVERSAL DECLARATION OF HUMAN RIGHTS as a common standard of achievement for all peoples and all nations, to the end that every individual and every organ of society, keeping this Declaration constantly in mind, shall strive by teaching and education to promote respect for these rights and freedoms and by progressive measures, national and international, to secure their universal and effective recognition and observance, both among the peoples of Member States themselves and among the peoples of territories under their jurisdiction.

Article 1

All human beings are born free and equal in dignity and rights.They are endowed with reason and conscience and should act towards one another in a spirit of brotherhood.

Article 2

Everyone is entitled to all the rights and freedoms set forth in this Declaration, without distinction of any kind, such as race, colour, sex, language, religion, political or other opinion, national or social origin, property, birth or other status. Furthermore, no distinction shall be made on the basis of the political, jurisdictional or international status of the country or territory to which a person belongs, whether it be independent, trust, non-self-governing or under any other limitation of sovereignty.

Article 3

Everyone has the right to life, liberty and security of person.

Article 4

No one shall be held in slavery or servitude; slavery and the slave trade shall be prohibited in all their forms.

Article 5

No one shall be subjected to torture or to cruel, inhuman or degrading treatment or punishment.

Article 6

Everyone has the right to recognition everywhere as a person before the law.

Article 7

All are equal before the law and are entitled without any discrimination to equal protection of the law. All are entitled to equal protection against any discrimination in violation of this Declaration and against any incitement to such discrimination.

Article 8

Everyone has the right to an effective remedy by the competent national tribunals for acts violating the fundamental rights granted him by the constitution or by law.

Article 9

No one shall be subjected to arbitrary arrest, detention or exile.

Article 10

Everyone is entitled in full equality to a fair and public hearing by an independent and impartial tribunal, in the determination of his rights and obligations and of any criminal charge against him.

Article 11

(1) Everyone charged with a penal offence has the right to be presumed innocent until proved guilty according to law in a public trial at which he has had all the guarantees necessary for his defence.

(2) No one shall be held guilty of any penal offence on account of any act or omission which did not constitute a penal offence, under national or international law, at the time when it was committed. Nor shall a heavier penalty be imposed than the one that was applicable at the time the penal offence was committed.

Article 12

No one shall be subjected to arbitrary interference with his privacy, family, home or correspondence, nor to attacks upon his honour and reputation. Everyone has the right to the protection of the law against such interference or attacks.

Article 13

(1) Everyone has the right to freedom of movement and residence within the borders of each state.

(2) Everyone has the right to leave any country, including his own, and to return to his country.

Article 14

(1) Everyone has the right to seek and to enjoy in other countries asylum from persecution.

(2) This right may not be invoked in the case of prosecutions genuinely arising from non-political crimes or from acts contrary to the purposes and principles of the United Nations.

Article 15

(1) Everyone has the right to a nationality.

(2) No one shall be arbitrarily deprived of his nationality nor denied the right to change his nationality.

Article 16

(1) Men and women of full age, without any limitation due to race, nationality or religion, have the right to marry and to found a family. They are entitled to equal rights as to marriage, during marriage and at its dissolution.

(2) Marriage shall be entered into only with the free and full consent of the intending spouses.

(3) The family is the natural and fundamental group unit of society and is entitled to protection by society and the State.

Article 17

(1) Everyone has the right to own property alone as well as in association with others.

(2) No one shall be arbitrarily deprived of his property.

Article 18

Everyone has the right to freedom of thought, conscience and religion; this right includes freedom to change his religion or belief, and freedom, either alone or in community with others and in public or private, to manifest his religion or belief in teaching, practice, worship and observance.

Article 19

Everyone has the right to freedom of opinion and expression; this right includes freedom to hold opinions without interference and to seek, receive and impart information and ideas through any media and regardless of frontiers.

Article 20

(1) Everyone has the right to freedom of peaceful assembly and association.

(2) No one may be compelled to belong to an association.

Article 21

(1) Everyone has the right to take part in the government of his country, directly or through freely chosen representatives.

(2) Everyone has the right of equal access to public service in his country.

(3) The will of the people shall be the basis of the authority of government; this will shall be expressed in periodic and genuine elections which shall be by universal and equal suffrage and shall be held by secret vote or by equivalent free voting procedures.

Article 22

Everyone, as a member of society, has the right to social security and is entitled to realization, through national effort and international co-operation and in accordance with the organization and resources of each State, of the economic, social and cultural rights indispensable for his dignity and the free development of his personality.

Article 23

(1) Everyone has the right to work, to free choice of employment, to just and favourable conditions of work and to protection against unemployment.

(2) Everyone, without any discrimination, has the right to equal pay for equal work.

(3) Everyone who works has the right to just and favourable remuneration ensuring for himself and his family an existence worthy of human dignity, and supplemented, if necessary, by other means of social protection.

(4) Everyone has the right to form and to join trade unions for the protection of his interests.

Article 24

Everyone has the right to rest and leisure, including reasonable limitation of working hours and periodic holidays with pay.

Article 25

(1) Everyone has the right to a standard of living adequate for the health and well-being of himself and of his family, including food, clothing, housing and medical care and necessary social services, and the right to security in the event of unemployment, sickness,

disability, widowhood, old age or other lack of livelihood in circumstances beyond his control.

(2) Motherhood and childhood are entitled to special care and assistance. All children, whether born in or out of wedlock, shall enjoy the same social protection.

Article 26

(1) Everyone has the right to education. Education shall be free, at least in the elementary and fundamental stages. Elementary education shall be compulsory. Technical and professional education shall be made generally available and higher education shall be equally accessible to all on the basis of merit.

(2) Education shall be directed to the full development of the human personality and to the strengthening of respect for human rights and fundamental freedoms. It shall promote understanding, tolerance and friendship among all nations, racial or religious groups, and shall further the activities of the United Nations for the maintenance of peace.

(3) Parents have a prior right to choose the kind of education that shall be given to their children.

Article 27

(1) Everyone has the right freely to participate in the cultural life of the community, to enjoy the arts and to share in scientific advancement and its benefits.

(2) Everyone has the right to the protection of the moral and material interests resulting from any scientific, literary or artistic production of which he is the author.

Article 28

Everyone is entitled to a social and international order in which the rights and freedoms set forth in this Declaration can be fully realized.

Article 29

(1) Everyone has duties to the community in which alone the free and full development of his personality is possible.

(2) In the exercise of his rights and freedoms, everyone shall be subject only to such limitations as are determined by law solely for the purpose of securing due recognition and respect for the rights and freedoms of others and of meeting the just requirements of morality, public order and the general welfare in a democratic society.

(3) These rights and freedoms may in no case be exercised contrary to the purposes and principles of the United Nations.

Article 30

Nothing in this Declaration may be interpreted as implying for any State, group or person any right to engage in any activity or to perform any act aimed at the destruction of any of the rights and freedoms set forth herein.

5.10 Child Welfare

The Indian Government, with its various schemes, caters to all the citizens of India irrespective of their caste, creed, gender, age and geographical origin. Children, often referred to as the leaders of tomorrow, are taken due care of by the government by various means and measures. According to Article 39 of the Directive Principles of State Policy, the Constitution of India pledges that the State will direct its policy towards children, with an aim of providing them the opportunities and facilities to grow in a healthy manner, in conditions of freedom and dignity, and to protect them from exploitation.

Child welfare programmes have occupied a prominent place in the national plans of the human resource development of the government. In this regard, India has adopted the National Policy on Children in 1974. It is aimed at ensuring equality of opportunity to the children. The policy provides the framework to address the needs of the children.

India is a signatory to the World Declaration on the Survival, Protection and Development of Children. In pursuance of its commitment, the Department of Women and Child Development has formulated the National Plan of Action for Children. The areas addressed by the plan include health, nutrition, education, water, sanitation and environment. The nation has also addressed the polio disease by providing free vaccination camps for children to be vaccinated against polioThe female child is often treated as the source of prosperity of the house. However, there have been cases where the female child has been considered a burden by her parents. To ensure that the female child is no longer discriminated, the government has been campaigning effectively. The slogan of the Indian government to caimpaign for the Girl Child has been A Happy Girl is the Future of our Country. The National Plan of Action for the Girl Child, seeks to prevent female foeticide and infanticide, to eliminate discrimination based on gender, rehabilitate and protect girls from exploitation, assault and abuse

The Education for All campaign of the government addresses 19 to 24 million children in the age group 6 to 14, of which 60 percent are girls. Apart from these activities, numerous Non-Governmental Organizations in India are providing means to provide shelter, better health, education and training to the street children, thereby rehabilitating them.

India's stand on child labour has evolved over years. The Child Labour Prohibition and Regulation Act of the government aims at banning employment of children below the age of fourteen years in factories, mines and hazardous employment.

5.11 Women Welfare

Since India gained Independence in 1948, the government has been focussing on the welfare and progress of women. All the five-year development plans of India included a special section on women, children and their development. From an initial plan outlay of Rs.4 crores to Rs.2000 crores, the government has made efforts to bring up the status of women socially, economically and politically on par with that of men. Education is now compulsory for women. In the State of Gujarat, high school and college education is subsidised for women. At present, the Women's Bill which holds 33% reservation for women's representatives in the parliament is awaiting approval in the lower house.

Among the many policies launched by the government are those related to improving the status of women in India. As per the Child Marriage Restraint Act, it is illegal for a girl child under the age of 18 to be married. The Maternity Benefits Act focuses on the child and mother health care. The Dowry Prohibition Act is aimed at abolishing the system of dowry (the giving of gifts to the groom by the bride's family as per the groom's family demands). The Family Courts are targetted towards providing women with legal aid and settling family related issues.

The government of India has set up forums and drawn up action plans for the development of women. The National Plan of Action for the Girl Child ensures the survival, protection and development of the girl child. The

National Nutritional Policy concentrates on improving the nutritional status of women and children. The Shram Shakthi report has recommended areas related to improving employment opportunities, training and development of skills for women.

The government of India has set up forums and drawn up action plans for the development of women. The National Plan of Action for the Girl Child ensures the survival, protection and development of the girl child. The National Nutritional Policy concentrates on improving the nutritional status of women and children. The Shram Shakthi report has recommended areas related to improving employment opportunities, training and development of skills for women.

Test Your Understanding

1. What is population explosion? Explain
2. Explain the causes of rapid population growth
3. What are the various ways and means to solve over population?
4. What is family welfare Program? And Explain the effects of its achievements in various Five Year plan
5. Explain the various features of Human rights adapted in General assembly of United Nations
6. Write short notes on
   (i) Child Welfare
   (ii) Women welfare

ANNEXURE I

The Air (Prevention and Control of Pollution) Act, 1981

CHAPTER I. PRELIMINARY

1. Short Title, Extent and Commencement.

(1) This Act may be called the Air (Prevention and Control of Pollution) Act, 1981.

(2) It extends to the whole of India.

(3) It shall come into force on such date as the Central Government may, by notification in the Official Gazette, appoint.

2. Definitions

In this Act, unless the context otherwise requires,-

(a) "air pollutant" means any solid, liquid or gaseous substance (including noise) present in the atmosphere in such concentration as may be or tend to be injurious to human beings or other living creatures or plants or property or environment;

(b) "air pollution" means the presence in the atmosphere of any air

(c) "approved appliances" means any equipment or gadget used for the bringing of any combustible material or for generating or consuming any fume, gas of particulate matter and approved by the State Board for the purpose of this Act;

(d) "approved fuel" means any fuel approved by the State Board for the purposes of this Act;

(e) "automobile" means any vehicle powered either by internal combustion engine or by any method of generating power to drive such vehicle by burning fuel;

(f) "Board" means the Central Board or State Board;

(g) "Central Board- means the [Central Board for the Prevention and Control of Water Pollution] constituted under section 3 of the Water (Prevention and Control of Pollution) Act, 1974;

(h) "chimney" includes any structure with an opening or outlet from or through which any air pollutant may be emitted,

(i) "control equipment" means any apparatus, device, equipment or system to control the quality and manner of emission of any air pollutant and includes any device used for securing the efficient operation of any industrial plant;

(j) "emission" means any solid or liquid or gaseous substance coming out of any chimney, duct or flue or any other outlet;

(k) "industrial plant" means any plant used for any industrial or trade purposes and emitting any air pollutant into the atmosphere;

(l) "member" means a member of the Central Board or a State Board, as the case may be, and includes the Chairman thereof,

(m) "occupier", in relation to any factory or premises, means the person who has control over the affairs of the factory or the premises, and includes, in relation to any substance, the person in posse ssion of the substance;

(n) "prescribed" means prescribed by rules made under this Act by the Central Government or as the case may be, the State government;

(o) "State Board" means,-

(i) in relation to a State in which the Water (Prevention and Control of Pollution) Act, 1974, is in force and the State Government has constituted for that State a [State Board for the Prevention and Control of Water Pollution] under section 4 of that Act, the said State Board; and

(ii) in relation to any other State, the State Board for the Prevention and Control of Air Pollution constituted by the State Government under section 5 of this Act.

CHAPTER II : CENTRAL AND STATE BOARDS FOR THE PREVENTION AND CONTROL OF AIR POLLUTION

3. [Central Board for the Prevention and Control of Air Pollution

The Central Board for the Prevention and Control of Water Pollution constituted under section 3 of the Water (Prevention and Control of Pollution) Act, 1974 (6 of 1974), shall, without prejudice to the exercise and performance of its powers and functions under this Act, exercise the powers and perform the functions of the Central Board for the Prevention and Control of Air Pollution under this Act.

4. [State Boards for the Prevention and Control of Water Pollution to be, State Boards for the Prevention and Control of Air Pollution

In any State in which the Water (Prevention and Control of Pollution) Act, 1974 (6 of 1974), is in force and the State Government has constituted for that State a State Board for the Prevention and Control of Water Pollution under section 4 of that Act, such State Board shall be deemed to be the State Board for the Prevention and Control of air Pollution constituted under section 5 of this Act and accordingly that State Board for the Prevention and Control of Water Pollution shall, without prejudice to the exercise and performance of its powers and functions under that Act, exercise the powers and perform the functions of the State Board for the Prevention and Control of Air Pollution under this Act.

5. Constitution of State Boards

(1) In any State in which the Water (Prevention and Control of Pollution) Act, 1974 (6 of 1974), is not in force, or that Act is in force but the State Government has not constituted a [State Board for the Prevention and Control of Water Pollution] under that Act, the State Government shall, with effect from such date as it may, by notification in the Official Gazette, appoint, constitute a State Board for the Prevention and Control of Air Pollution under such name as may be specified in the notification, to exercise the powers conferred on, and perform the functions assigned to, that Board under this Act.

(2) A State Board constituted under this Act shall consist of the following members, namely:-

(a) a Chairman, being a person, having a person having special knowledge or practical experience in respect of matters relating to environmental protection, to be nominated by the State Government:

Provided that the Chairman my be either whole-time or part-time as the State Government may think fit;

(b) such number of officials, not exceeding five, as the State Government may think fit, to be nominated by the State Government to represent that government;

(c) such number of persons, not exceeding five, as the State Government may think fit, to be nominated by the State Government from amongst the members of the local authorities functioning within the State;

(d) such number of non-officials, not exceeding three, as the State Government may think fit, to be nominated by the State Government to represent the interest of agriculture, fishery or industry or trade or labour or any other interest, which in the opinion of that government, ought to be represented;

(e) two persons to represent the companies or corporations owned, controlled or managed by the State Government, to be nominated by that Government;

(f) a full-time member-secretary having such qualifications knowledge and experience of scientific, engineering or management aspects of pollution control as may be prescribed, to be appointed by the State Governments;

Provided that the State Government shall ensure that not less than two of the members are persons having special knowledge or practical experience in, respect of matters relating to the improvement of the quality of air or the prevention, control or abatement of air pollution.

(3) Every State Board constituted under this Act shall be a body corporate with the name specified by the State Government in the notification issued under sub-section (1), having perpetual succession and a common seal with power, subject to the provisions of this Act, to acquire and dispose of property and to contract, and may by the said name sue or be sued.

6. Central Board to Exercise the Powers and Perform Die Functions of a State Board in the Union Territories

No State Board shall be constituted for a Union territory and in relation to -a Union territory, the Central Board shall exercise the powers and perform the functions of a State Board under this Act for that Union territory

Provided that in relation to any Union territory the Central Board may delegate all or any of its powers and functions under this section to such person or body of persons as the Central Government may specify.

7. Terms and Conditions of Service of Members

(1) Save as otherwise provided by or under this Act, a member of a State Board constituted under this Act, other than the member-secretary, shall hold office for a term of three years from the date on which his nomination is notified in the Official Gazette:

Provided that a member shall, notwithstanding the expiration of his term, continue to hold office until his successor enters upon his office.

(2) The terms of office of a member of a State Board constituted under this Act and nominated under clause (b) or clause *(e)* of sub-section (2) of section 5 shall come to an end as soon as he ceases to hold the office under the State Government as the case may be, the company or corporation owned, controlled

or managed by the State Government, by virtue of which he was nominated.

(3) A member of a State Board constituted under this Act, other than the member- secretary, may at any time resign his office by writing under his hand addressed,-

(a) in the case of the Chairman, to the State Government; and

(b) in any other case, to the Chairman of the State Board, and the seat of be Chairman or such other member shall thereupon become vacant.

(4) A member of a State Board constituted under this Act, other than the member-secretary, shall be deemed to have vacated his scat, if he is absent without reason, sufficient in the opinion of the State Board, from three consecutive meetings of the State Board or where he is nominated under clause (c) of subsection (2) of section 5, he ceases to be a member of the local authority and such vacation of scat shall, in either case, take effect from such as the State Government may, by notification in the Official Gazette, specify.

(5) A casual vacancy in a State Board constituted under this Act shall be filled by a fresh nomination and the person nominated to fill the vacancy shall hold office only for the remainder of die term for which the member whose place lie takes was nominated.

(6) A member of a State Board constituted under this Act shall be eligible for re-nomination

(7) The other terms and conditions of service of the Chairman and other members (except the member-secretary) of a State Board constituted under this Act shall be such as may be prescribed.

8. Disqualifications

(1) No person shall be a member of a State Board constituted under this

(a) is, or at any time has been, adjudged insolvent, or

(b) is of unsound mind and has been so declared by a competent court,

(c) is, or has been, convicted *of* an offence which, in the opinion of the State Government, involves moral turpitude, or

(d) is, or at any time has been, convicted of an offence under this Act,

(e) has directly or indirectly by himself on by any partner.. any share or interest in any Finn or company carrying on the business of manufacture, sale, or hire of machinery, industrial plant, c6ntrol equipment or any other apparatus for the improvement of the quality of air or for the prevention, control or abatement of air pollution, or

(f) is a director or a secretary, manager or other salaried officer or employee of any company or firm having any contract with the Board, or with the Government constituting the Board or with a local authority in the State, or with a company or corporation owned, controlled or managed by the Government, for the carrying out of programmes for the improvement of the quality of air or for the prevention, control or abatement of air pollution, or

(g) has so abused, in the opinion of the State Government, his position as a member, as to render his continuance on the State Board detrimental to the interest of the general public.

(2) The State Government shall, by order in writing, remove any member who is, or has become, subject to any disqualification mentioned in sub-section M.

Provided that no order of removal shall be made by the State Government under this section unless the member concerned has been given a reasonable opportunity of showing cause against the same.

(3) Notwithstanding anything contained in sub-section (1) or sub-section (6) of section 7, a member who has been removed under this section shall not be eligible to continue to hold office until his successor enters upon his office, or, as the case may be, for re-nomination as a member.

9. Vacation of Seats by Members

If a member of a State Board constituted under this Act becomes subject to any of the disqualifications specified in section 8, his seat shall become vacant.

10. Meetings-of Board

(1) For the purposes of this Act, a Board shall meet at least once in every three months and shall observe such rules of procedure in regard to the transaction of business at its meetings as may be prescribed:

Provided that it, in the opinion of the Chairman, any business of an urgent nature is to be transacted, he may convene a meeting of the Board at such time as he thinks fit for the aforesaid purpose.

(2) Copies of minutes of the meetings under sub-section (1) shall be forwarded to the Central Board and to the State Government concerned.

11. Constitution -of Committees

(1) A Board may constitute as many committees consisting wholly of members or partly of members and partly of other persons and for such purpose or purposes as it may think fit.

(2) A committee constituted under this section shall meet at such time and at such place, and shall observe such rules of procedure in regard to the transaction of business at its meetings, as may be prescribed.

(3) The members of a committee other than the members of the Board shall be paid such fees and allowances, for attending its meetings and for attending to any other work of the Board as may be prescribed.

12. Temporary Association of Persons with Board for Particular Purposes

(1) A Board may associate with itself in such manner, and for such purposes, as may be prescribed, any person whose assistance or advice it may desire to obtain in performing any of its functions under this Act.

(2) A person associated with the Board under sub-section (1) for any purpose shall have a right to take part in the discussions of the Board relevant to that purpose, but shall riot have a tight to vote at a meetings of the Board and shall not be a member of the Board for any other purpose.

(3) A person associated with a Board under sub-section (1) shall be entitled to receive such fees and allowances as may be prescribed.

13. Vacancy in Board not to Invalidate Acts or Proceedings

No act or proceeding of a Board or any committee thereof shall be called in question on the ground merely of the existence of any vacancy in or any defect in the constitution of, the Board or such committee, as the case may be.

14. Member-Secretary and Officers and Other Employees of State Boards

(1) The terms and conditions of service of the member-secretary of a State Board constituted under this Act shall be such as may be prescribed.

(2) The member-secretary of a State Board, whether constituted under this Act or not, shall exercise such powers and perform such duties as may be prescribed or as may, from time to time, be delegated to him by the State Board or its Chairman.

(3) Subject to such rules as may be made by the State Government in this behalf, a State Board, whether constituted under this Act or not, may appoint such officers and other employees as it considers necessary for the efficient performance of its functions under this Act.

(4) The method of appointment, the conditions of service and the scale of pay of the officers (other than the member-secretary) and other employees of a State Board appointed under sub-section (3) shall be such as may be determined by regulations made by the State Board under this Act.

(5) Subject to such conditions as may be prescribed, a State Board constituted under this Act may from time to time appoint any qualified person to be a consultant to the Board and pay him such salary and allowances or fees, as it thinks fit.

15. Delegation of Powers

A State Board may, by general or special order, delegate to t1he Chairman or the member-secretary or any other officer of the Board subject to such conditions and limitations, if any. as may be specified in the order, such of its powers and functions under this Act as It may deem necessary.

CHAPTER III : POWERS AND FUNCTIONS OF BOARDS

16. Functions of Central Board

(1) Subject to the provisions of this Act, and without prejudice to the performance, of its functions under the Water (Prevention and Control of Pollution) Act, IL974 (6 of 1974), the main functions of the Central Board shall be to improve the quality of air and to prevent, control or abate air pollution in the country.

(2) In particular and without prejudice to the generality of the foregoing functions, the Central Board may-

(a) advise the Central Government on any matter concerning the improvement of the quality of air and the prevention, control or abatement of air pollution;

(b) plan and cause to be executed a nation-wide programme for the prevention, control or abatement of air pollution;

(c) co-ordinate the activities of the State and resolve disputes among them;

(d) provide technical assistance and guidance to the State Boards, carry out and sponsor investigations and research relating to problems of air pollution and prevention, control or abatement of air pollution;

[(i) perform such of the function of any State Board as may, be specified in and order made under sub-section (2) of section 18;]

(e) plan and organise the training of persons engaged or to be engaged in programmes for the prevention, control or abatement of air pollution on such terms and conditions as the Central Board may specify;

(f) organise through mass media a comprehensive programme regarding the prevention, control or abatement of air pollution;

(g) collect, compile and publish technical and statistical data relating to air pollution and the measures devised for its effective prevention, control or abatement and prepare manuals, codes or guides relating to prevention, control or abatement of air pollution;

(h) lay down standards for the quality of air.,

(i) collect and disseminate information in respect of matters relating to air pollution;

(j) perform such other functions as may be prescribed.

(3) The Central Board may establish or recognise a laboratory or laboratories to enable the Central Board to perform its functions under this section efficiently.

(4) The Central Board may-

(a) delegate any of its functions under this Act generally or specially to any of the committees appointed by it;

(b) do such other things and perform such other acts as it may think necessary for the proper discharge of its functions and generally for the purpose of carrying into effect the purposes of this Act.

17. Functions of State Boards

(1) Subject to the provisions of this Act, and without prejudice to the performance of its functions, if any, under the Water (Prevention and Control of Pollution) Act, 1974 (Act 6 of 1974), the functions of a State Board shall be-

(a) to plan a comprehensive programme for the prevention, control or abatement of air pollution and to secure the execution thereof-,

(b) to advise the State Government on any matter concerning the prevention, control or abatement of air pollution;

(c) to collect and disseminate information relating to air pollution;

(d) to collaborate with the Central Board in organising the training of persons engaged or to be engaged in programmes relating to prevention, control or abatement of air pollution and to organise mass-education programme relating thereto;

(e) to inspect, at all reasonable times, any control equipment, industrial plant or manufacturing process and to give, by order, such directions to such persons as it may consider necessary to take steps for the prevention, control or abatement of air pollution;

(f) to inspect air pollution control areas at such intervals as it may think necessary, assess the quality of air therein and take steps for the prevention, control or abatement of air pollution in such areas;

(g) to lay down, in consultation with the Central Board and having regard to the standards for the quality of air laid down by the Central Board, standards for emission of air pollutants into the atmosphere from industrial plants and automobiles or for the discharge of any air pollutant into the atmosphere from any other source whatsoever not being a ship or an aircraft:

Provided that different standards for emission may be laid down under this clause for different industrial plants having regard to the quantity and composition of emission of air pollutants into the atmosphere from such industrial plants;

(h) to advise the State Government with respect to the suitability of any premises or location for carrying on any industry which is likely to cause air pollution;

(i) to Perform such other functions as may be prescribed or as may, from time to time, be entrusted to it by the Central Board or the State Government;

(j) to do such other things and to perform such other acts as it may think necessary for the proper discharge of its functions and generally for the purpose of carrying into effect the purposes of this Act.

(2) A State Board may establish or recognise a laboratory or laboratories to enable the State Board to perform its functions under this section efficiently.

18. Power to Give Directions

(1) [In the performance of its functions under this Act-

(a) the Central Board shall be bound by such directions in writing as the Central Government may give to it; and

(b) every State Board shall be bound by such directions in writing as the Central Board or the State Government may give to it:

Provided that where a direction given by the State Government is inconsistent with the direction given by the Central Board, the matter shall be referred to the Central Government for its decision.

(2) [Where the Central Government is of the opinion that any State Board has defaulted in complying with any directions given by the Central Board under sub-section (1) and as a result of such default a grave emergency has arisen and it is necessary or expedient so to do in the public interest, it m4y, by order, direct the Central Board to perform any of the functions of the State Board in relation to such area, for such period and for such purposes, as may be specified in the order.

(3) Where the Central Board performs any of the functions of the State Board in pursuance of a direction under sub-section (2), the expenses, if any incurred by the Central Board with respect to the performance of such functions may, if the State Board is empowered to recover such expenses, be recovered by the Central Board with interest (at such reasonable rate as the Central Government may, by order, fix) from the date when a demand for such expenses is made until it is paid from the person or persons concerned as arrears of land revenue or of public demand.

(4) For the removal of doubts, it is hereby declared that any directions to perform the functions of any State Board given under sub-section (2) in respect of any area would not preclude the State Board from performing such functions in any other area in the State or any of its other functions'in that area.

CHAPTER IV : PREVENTION AND CONTROL OF AIR POLLUTION

19. Power to Declare Air Pollution Control Areas

(1) The State Government may, after consultation with the State Board, by notification in the Official Gazette declare in such manner as may be prescribed, any area or areas within the State as air pollution control area or areas for the purposes of this Act.

(2) The State government may, after consultation with the State Board, by notification in the Official Gazette,-

(a) alter any air pollution control area whether by way of extension or reduction ;

(b) declare a new air pollution control area in which may be merged one or more existing air pollution control areas or any part or parts thereof.

(3) If the State Government, after consultation with the State Board, is of opinion that the use of any fuel, other than an approved fuel, in any air pollution control area or part thereof, may cause or is likely to cause air pollution, it may, by notification in the Official Gazette, prohibit the use of such fuel in such area or part thereof with effect from such date (being not less than three months from the date of publication of the notification) as may be specified in the notification.

(4) The State Government may, after consultation with the Sate Board, by notification in the Official Gazette, direct that with effect from such date as may be specified therein, no appliance, other than an approved appliance, shall be used in the premises situated in an air pollution control area :

Provided that different dates may be specified for different parts of an air pollution control area or for the use of different appliances.

(5) If the State Government, after consultation with the State Board, is of opinion that the burning of any material (not being fuel) in any air pollution control area or part thereof may cause or is likely to cause air pollution, it may, by notification in the Official Gazette, prohibit the burning of such material in such area or part thereof.

20. Power to Give Instructions for Ensuring Standards for Emission from Automobiles

With a view to ensuring that the standards for emission of air pollutants from automobiles laid down by the State Board tinder clause (g) of sub-section (1) of section 17 are complied with, the State Government shall, in consultation with the State Board, give such instructions as may be deemed necessary to the concerned authority in charge of registration of motor vehicles under the Motor Vehicles Act, 1939 (Act 4 of 1939), and such authority shall, notwithstanding anything contained in that Act or the rules made thereunder be bound to comply with such instructions.

21. Restrictions on Use of Certain Industrial Plants

(1) Subject to the provisions of this section, no person shall, without the previous consent of the State Board, establish or operate any industrial plant in an air pollution control area :

Provided that a person operating any industrial plant in any air pollution control area, immediately before the commencement of section 9 of the Air (Prevention and Control of Pollution) Amendment Act, 1987, for which no consent was necessary prior to such commencement, may continue to do so for a period of three months from such commencement or, if he has made an application for such consent within the said period of three months, till the disposal of such application.

(2) An application for consent of the State Board under sub-section (1) shall be accompanied by such fees as may bc prescribed 'and shall be made in the prescribed form and shall contain the particulars of the industrial plant and such other particulars as may be prescribed :

Provided that where any person, immediately before the declaration of any area as an air pollution control area, operates in such area any industrial plant,[16] such person shall make the application under this sub-section within such period (being not less than three months from the date of such declaration) as may be prescribed and where such person makes such application, he shall be deemed to be operating such industrial plant with the consent of the State Board until the consent applied for has been refused,

(3) The State Board may make such inquiry as it may deem fit in respect of the application for consent referred to in sub-section (1) and in making any such inquiry, shall follow such procedure as may be prescribed.

(4) Within a period of four months after the receipt of the application for consent referred to in sub-section (1), the State Board shall, by order in writing, and for reasons to be recorded in the order, grant the consent applied for subject to such conditions and for such period as may be specified in the order, or refuse consent:

Provided that it shall be open to the State Board to cancel such consent before the expiry of the period for which it is granted or refuse further consent after such expiry if the conditions subject to which such consent has been granted are not fulfilled:

Provided further that before cancelling a consent or refusing a further consent under the first provision, a reasonable opportunity of being heared shall be given to the person concerned.

(5) Every person to whom consent has been granted by the State Board under sub-section (4), shall comply with the following conditions, namely -

(i) the control equipment of such specifications as the State Board may approve in this behalf shall be installed and operated in the premises where the industry is carried on or proposed to be carried on;

(ii) the existing control equipment, if any, shall be altered or replaced in accordance with the directions of the State Board;

(iii) the control equipment referred to in clause (i) or clause (ii) shall be kept at all times in good running condition;

(iv) chimney, wherever necessary, of such specifications as the State Board may approve in this behalf shall be erected or re-erected in such premises; and

(v) such other conditions as the State Board, may specify in this behalf,

(vi) the conditions referred to in clauses (i), (ii) and (iv) shall be complied with within such period as the State Board may specify in this behalf-

Provided that in the case of a person operating any industrial plant in an air pollution control area immediately before the date of declaration of such area as an air pollution control area, the period so specified shall not be less than six months :

Provided further that-

(a) after the installation of any control equipment in accordance with the specifications under clause (i), or

(b) after the alteration or replacement of any control equipment in accordance with the directions of the State Board under clause (ii), or

(c) after the erection or re-erection of any chimney under clause (iv), no control equipment or chimney shall be altered or replaced or, as the case may be, erected or re-created except with the previous approval of the State Board.

(6) If due to any technological improvement or otherwise the State Board is of opinion that all or any of the conditions referred to in sub-section (5) require or requires variation (including the change of any control equipment, either in whole or in part), the State Board shall, after giving the person to whom consent has been granted an opportunity of being heard, vary all or any of such conditions and thereupon such person shall be bound to comply with the conditions as so varied.

(7) Where a person to whom consent has been granted by the State Board under sub-section (4) transfers his interest in the industry to any other person, such consent shall be deemed to have been granted to such other person and he shall be bound to comply with all the conditions subject to which it was granted as if the consent was granted to him originally.

22. Persons carrying on Industry, etc., and to Allow Emission of Air Pollutants in Excess of the Standard Laid Down by State Board.

No person operating any industrial plant, in any air pollution control area shall discharge or cause or permit to be discharged the emission of any air pollutant in excess of the standards laid down by the State Board under clause (g) of sub-section (1) of section 17.

22A. Power of Board to Make Application to Court for Restraining Person from Causing Air Pollution

(1) Where it is apprehended by a Board that emission of any air pollutant, in excess of the standards laid down by the State Board under clause (g) of sub-section (1) of section 17, is likely to occur by reason of any person operating an industrial plant or otherwise in any air pollution control area, the Board may make an application to a court, not inferior to that of a Metropolitan Magistrate or a Judicial Magistrate of the first class for restraining such person from emitting such air pollutant.

(2) On receipt of the application under sub-section (1), the court may make such order as it deems fit.

(3) Where under sub-section (2), the court makes an order restraining any person from discharging or causing or permitting to be discharged the emission of any air pollutant, it may, in that order,-

(a) direct such person to desist from taking such action as is likely to cause emission;

(b) authorise the Board, if the direction under clause (a) is not complied with by the person to whom such direction is issued, to implement the direction in such manner as may be specified by the court.

(4) All expenses incurred by the Board in implementing the sections of the court under clause (b) of sub-section (3) sl)all be recoverable from the person concerned as an-ears of land revenue or of public demand.

23. Furnishing of Information to State Board and Other Agencies in Certain Cases

(1) Where in any area the emission of any air pollutant into the atmosphere in excess of the standards laid down by the State Board occurs or is apprehended to occur due to accident or other unforeseen act or event, the person in charge of the premises from where which emission occurs or is apprehended to occur shall forthwith intimate the fact of such occurrence or the apprehension of such occurrence to the State Board and to such authorities or agencies as may be prescribed.

(2) On receipt of information with respect to the fact or the apprehension of any occurrence of the nature referred to in sub-section (1) whether through intimation under that sub-section or otherwise, the State Board and the authorities or agencies shall, as early as practicable, cause such remedial measure to be I aken as are necessary to mitigate the emission of such air pollutants.

(3) Expenses, if any, incurred by the State Board, authority or agency with respect to the remedial measures referred to in sub-section (2) together with interest ("t such reasonable rate, as the State Government may, by order, fix) from the date when a demand for the expenses is made until it is paid, may be recovered by that Board, authority or agency from the person concerned, as arrears of land revenue, or of public demand.

24. Power of Entry and Inspection

(1) Subject to the provisions of this section, any person empowered by a State Board in this behalf shall have a right to enter, at all reasonable times with such assistance as he considers necessary, any place—

(a) for the purpose of performing any of the functions of the State Board entrusted to him :

(b) for the purpose of determining whether and if so in what manner, any such functions are to be performed or whether any provisions of this Act or the rules made thereunder or any notice, order, direction or authorisation served, made, given or granted under this Act is being or has been complied with;

(c) for the purpose of examining and testing any control equipment, industrial plant, record, register, document or any other material object or for conducting a search of any place in which he has reason to believe that an offence under this Act or the rules made has been or is being or is about to be committed and for seizing any such control equipment, industrial plant, record, register, document or other material object if he has reasons to believe that it may furnish evidence of the commission of an offence punishable under this Act or the rules made thereunder.

(2) Every person operating any control equipment or any industrial plant, in an air pollution control area shall be bound to render all assistance to the person empowered by the State Board under sub-section (1) for carrying out the functions under that sub-section and if he fails to do so without any reasonable cause or excuse, he shall be guilty of an offence under this Act.

(3) If any person willfully delays or obstructs any person empowered by the State Board under sub-section (1) in the discharge of his duties, he shall be guilty of an offence under this Act.

(4) The provisions of the Code of Criminal Procedure, 1973, or, in relation to the State of Jammu and Kashmir, or any area, in which that Code is not in force, the provisions of any corresponding law in force in that State or area, shall, so far as may be, apply to any search or seizure under this section as they apply to any search or seizure made under the authority of a warrant issued under section 94 of the said Code or, as the case may be, under the corresponding provisions of the said law.

25. Power to Obtain Information

For the purposes of carrying out the functions entrusted to it, the State Board or any officer empowered by it in Ns behalf may call for any information (including information regarding the types of air pollutants emitted into the atmosphere and the level of the emission of such air pollutants) from the occupier or any other person carrying oil any industry or operating any control equipment or industrial plant and for the purpose of verifying the correctness of such information, the State Board or such officer shall have the right to inspect the premises where such industry, control equipment or industrial plant is being carried on or operated.

26. Power to Take Samples of Air or Emission and Procedure to be Followed in Connection Therewith

(1) A State Board or any officer empowered by it in this behalf shall have power to take, for the purpose of analysis, samples of air or emission from any

chimney, flue or duct or any other outlet in such manner as may be prescribed.

(2) The result of any analysis of a sample of emission taken under subsection (1) shall not be admissible in evidence in any legal proceeding unless the provisions of sub-sections (3) and (4) are complied with.

(3) Subject to the provisions of sub-section (4), when a sample of emission is taken for analysis under sub-section (1), the person taking the sample shall-

(a) serve on the occupier or his agent, a notice, then and there, in such form as may be prescribed, of his intention to have it so analysed;

(b) in the presence of the occupier or his agent, collect a sample of emission for analysis;

(c) cause the sample to be placed in a container or containers which shall be marked and sealed and shall also be signed both by the person taking the sample and the occupier or his agent;

(d) send, without delay, the container to the laboratory established or recognised by the State Board under section 17 or, if a request in that behalf is made by the occupier or his agent when the notice is served on him under clause (a), to the laboratory established or specified under sub-section (1) of section 28.

(4) When a sample of emission is taken for analysis under sub-section (1) and the person taking the sample serves on the occupier or his agent, a notice under clause (a) of sub-section (3), then,-

(a) in a case where the occupier or his agent willfully absents himself, the person taking the sample shall collect the sample of emission for analysis to be placed in a container or containers which shall be marked and sealed and shall also be signed by the person taking the sample, and

(b) in a case where the occupier or his agent is present at the time of taking the sample but refuses to sign the marked and scaled container or containers of the sample of emission as required under clause (c) of subsection (3), the marked and sealed container or containers shall be signed by the person taking the sample, and the container or containers shall be sent without delay by the person 'Caking the **sample for analysis** to the laboratory established or specified under sub-section (7) of section 28 and such person shall inform the Government analyst appointed under sub-section (1) of section 29, in writing, about the wilfull absence of the occupier or his agent, or, as the case may be, his refusal to sing the container or containers.

27. Reports of the Result of Analysis on Samples Taken Under Section 26

(1) Where a sample of emission has been sent for analysis to the laboratory established or recognised by the State Board, the Board analyst appointed under sub-section (2) of section 29 shall analyse the sample and submit a report in the prescribed form of such analysis in triplicate to the State Board.

(2) On receipt of the report under sub-section (1), one copy of the report shall be sent by the State Board to the occupier or his agent referred to in section 26, another copy shall be preserved for production before the court in case any legal proceedings are taken against him and the other copy shall be kept by the State Board.

(3) Where a sample has been sent for analysis under clause (a~ of sub-section (3) or sub-section (4) of section 26 to any laboratory mentioned therein, the Government analyst referred to in the said sub-section (4) shall analyse the sample and submit a report in the prescribed form of the result of the analysis in triplicate to the State Board which shall comply with the provisions of sub-section (2).

(4) Any cost incurred in getting any sample analysed at the request of the occupier or his agent as provided in clause (d) of sub-section (3) of section 26 or when he wilfully absents himself or refuses to sing the marked and scaled container or containers of sample of emission under sub-section (4) of that section, shall be payable by such occupier or his agent and in case of default the same shall be recoverable from him as arrears of land revenue or of public demand.

28. State Air Laboratory

(1) The State Government may, by notification in the Official Gazette,-

(a) establish one or more State Air Laboratories; or

(b) specify one or more laboratories or institutes as State Air Laboratories to carry out the functions entrusted to the State Air Laboratory under this Act.

(2) The State Government may, after consultation with the State Board, make rules prescribing-

(a) the functions of the State Air Laboratory;

(b) the procedure for the submission to the said Laboratory of samples of air or emission for analysis or tests, the form of the Laboratory's report thereon and the fees payable in respect of such report;

(c) such other matters as may be necessary or expedient to enable that Laboratory to carry out its functions.

29. Analysis

(1) The State Government may, by notification in the Official Gazette, appoint such persons as it thinks fit and having the prescribed qualifications to be government analysts for the purpose of analysis of samples of air or emission sent for analysis to any laboratory established or specified under sub-section (1) of section 28.

(2) Without prejudice to the provisions of section 14, the State Board may, by notification in the Official Gazette, and with the approval of the State Government, appoint such persons as it thinks fit and having the prescribed qualifications to be Board analysts for the purpose of analysis of samples of air or emission sent for analysis to any laboratory established or recognised under section 17.

30. Reports of Analysis

Any document purporting to be a report signed by a Government analyst or, as the case may be, a Statc Board analyst may be used as evidence of the facts stated therein in any proceeding under this Act.

31. Appeals

(1) Any person aggrieved by an order made by the State Board under this Act may, within thirty day from the date on which the order is communicated to him, prefer an appeal to such authority (hereinafter referred to as the Appellate Authority) as the State government may think fit to constitute :

Provided that the Appellate Authority may entertain the appeal after tile expiry of the said period of thirty days if such authority is satisfied that the appellant was prevented by sufficient cause from filing the appeal in time.

(2) The Appellate Authority shall consist of a single person or three persons as the State Government may think fit to be appoint by the State Government.

(3) The form and the manner in which an appeal may be preferred under subsection (1), the fees payable for such appeal and the procedure to be followed by the Appellate Authority shall be such as may be prescribed.

(4) On receipt of an appeal preferred under sub-section (1), the Appellate Authority shall, after giving the appellant and the State Board an opportunity of being heard, dispose of the appeal as expeditiously as possible.

31A. Power to Give Directions

Notwithstanding anything contained in any other law, im. subject to the provisions of this Act, and to any directions that the Central Government may give in this behalf, a Board may, in the exercise of its powers and performance of its functions under this Act, issue any directions in writing to any person, officer or authority, and such person, officer or authority shall be bound to comply with such directions.

Explanation: For the avoidance of doubts, it is hereby declared that tile power to issue directions under this section, includes the power to direct-

(a) the closure, prohibition or regulation of any industry, operation or

(b) the stoppage or regulation of supply of electricity, water or any other service.]

CHAPTER V : FUND, ACCOUNTS AND AUDIT

32. Contribution by Central Government

The Central Government may, after due appropriation made by Parliament by law in this behalf make in each financial year such contributions to the State Boards as it may think necessary to enable the State Board to perform their functions under this Act:

Provided that noting in this section shall apply to any [25][State Board for the Prevention and Control of water Pollution] constituted under section 4 of the Water (Prevention and Control of Pollution) Act, 1974, which is empowered by that Act to expend money from its fund thereunder also for. performing its functions, under any law for the time being in force relating to the prevention, control or abatement of air pollution.

33. Fund of Board

(1) Every State Board shall have its own fund for the purposes of this Act and all sums which may, from time to time, be paid to it by the *Central Government and all other receipts (by way of contributions, if any, from the State Government, fees, gifts, grants, donations benefactions or otherwise) of that Board shall be carried to the fund of the Board and all payments by the Board shall be made therefrom.

(2) Every State Board may expend such sums as it thinks fit for performing its functions under this Act and such sums shall be treated as expenditure payable out of the fund of that Board.

(3) Nothing in this section shall apply to any [State Board for the Prevention and Control of Water Pollution]

constituted under section 4 of the Water -(Prevention and Control of Pollution) Act, 1974, which is empowered by that Act to expend money from its fund thereunder also for performing its functions under any law for the time being in force relating to the prevention., control or abatement of air pollution.

33A. Borrowing Powers of Board

A Board may, with the consent of, or in accordance with the terms of any general or special authority given to it by, the Central Government or, as the case may be, the State Government, borrow money from any source by way of loans or issue of bonds, debentures or such other instruments, as it may deem fit, for discharging all or any of its functions under this Act.

34. Buduct

The Central Board or as the case may be the State Board shall, during each financial year, prepare, in such form and at such time as may be prescribed, a budget in respect of the financial year next ensuing showing the estimated receipt and expenditure under this Act, and copies thereof shall be forwarded to the Central Government or, as the case may be, the State Government.

35. Annual Report

(1) The Central Board shall, during each financial year, prepare, in such form as may be prescribed, an annual report giving full account of its activities under this Act during the previous financial year and copies thereof shall be forwarded to the Central Government within four months from the last date of the previous financial year and that Goverriment shall cause every such report to be laid before both Houses of Parliament within nine months of the last date of the previous financial year.

(2) Every State Board shall, during each financial year, prepare, in such fort-n as may be prescribed, an annual report giving full account of its activities under this Act during the previous financial year and copies thereof shall be forwarded to the State Government within four months from the last date of the previous financial year and that Government shall cause every such report to be laid before the State Legislature within a period of nine months from the date of the previous financial year.]

36. Accounts and Audit

(1) Every Board shall, in relation to its functions under this Act, maintain proper accounts and other relevant records and prepare an annual statement of accounts in such form as may be prescribed by the Central Government or, as the case may be, the State Government.

(2) The accounts of the Board shall be audited by an auditor duly qualified to act as an auditor of companies under section 226 of the Companies Act, 1956.

(3) The said auditor shall be appointed by the Central Government or, as the case may be, the State Government on the advice of the Comptroller and Auditor General of India.

(4) Every auditor appointed to audit the accounts of the Board under this Act shall have the right to demand the production of books, accounts, connected vouchers and other documents and papers and to inspect any of the offices of the Board.

(5) Every such auditor shall send a copy of his report together with an audited copy of the accounts to the Central Government or, as the case may be, the State Government.

(6) The Central Government shall, as soon as may be after the receipt of the audit report under sub-section (5), cause the same to be laid before both Houses of Parliament.

(7) The State Government shall, as soon as may be after the receipt of the audit report under sub-section (5), cause the same to be laid before the State Legislature.

CHAPTER VI. PENALTIES AND PROCEDURE

37. Failure to Comply with the Provisions of Section 21 or Section 22 or with the Directions Issued Under Section 31A.

(1) -whoever fails to comply with the provisions of section 21 or section 22 or directions issued under section 3 1 A, shall, in respect of each such failure, be punishable with imprisonment for a terms which shall not be less than one year and six months but which may extend to six years and with fine, and in case the failure continues, with an additional fine which may extend to five thousand rupees for every day during which such failure continues after the conviction for the first such failure.

(2) If the failure referred to in sub-section (1) continues beyond a period of one year after the date of conviction, the offender shall be punishable with imprisonment for a term which shall not be less than two years but which may extend to seven years and with fine.

38. Penalties for Certain Acts

Whoever-

(a) destroys, pulls down, removes, injures or defaces any pillar, post or stake fixed in the ground or any notice or other matter put up, incsribed or placed, by or under the authority of the Board, or

(b) obstructs any person acting under the orders or directions of the Board from exercising his powers and performing his functions under this Act, or

(c) damages any works or property belonging to the Board, or

(d) fails to furnish to the Board or any officer or other employee of the Board any information required by the Board or such officer or other employee for the purpose of this Act, or

(e) fails to intimate the occurrence of the emission of air pollutants into the atmosphere in excess of the standards laid down by the State Board or the apprehension of such occurrence, to the State Board and other prescribed authorities or agencies as required under sub-section (1) of section 23, or

(f) in giving any information which he is required to give under this Act, makes a statement which is false in any material particular, or

(g) for the purpose of obtaining any consent under section 21, makes a statement which is false in any material particular shall be punishable with imprisonment for a term which may extend to three months or with fine which may extend to [ten thousand rupees] or with both.

39. Penalty for Contravention of Provisions of the Act

Whoever contravenes any of the provisions of this Act or any order or direction issued thereunder, for which no penalty has been elsewhere provided in this Act, shall be punishable with imprisonment for a term which may extend to three months or with fine which may extend to ten thousand rupees or with both, and in the case of continuing contravention, with an additional fine which may extend to five thousand, rupees for every day during which such contravention continues after conviction for the first such contravention.

40. Offences by Companies

(1) Where an offence under this Act has, been committed by a company, every person who, at the time the offence was committed, was directly in charge of, and was responsible to, the company for the conduct of the business of the company, as well as the company, shall be deemed to be guilty of the offence and shall be liable to be proceeded against and punished accordingly:

Provided that nothing contained in this sub-section shall render any such person liable to any punishment provided in this Act, if he proves that the offence was committed without his knowledge or that he exercised all due diligence to prevent the commission of such offence.

(2) Notwithstanding anything contained in sub-section (1), where an offence under this Act has been committed by a company and it is proved that the offence has ben committed with the consent or connivance of, or is attributable to any neglect on the part of, any director, manager, secretary or other officer of the company, such director, manager, secretary or other officer shall also be deemed to be guilty of that offence and shall be liable to be proceeded against and punished accordingly.

Explanation: For the purpose of this section,-

(a) "company" means any body corporate, and includes a firin or other association of individuals; and

(b) "director", in relation to a firm, means a partner in the firm.

41. Offences by Government Departments

(1) Where an offence under this Act has been committed by any Department of Government, the Head of the Department shall be deemed to be guilty of the offence and shall be liable to be proceeded against and punished accordingly:

Provided that nothing contained in this section shall render such Head of the Department liable to any punishment if he proves that the offence was committed without his knowledge or that he exercised all due diligence to prevent the commission of such offence.

(2) Notwithstanding anything contained in sub-section (1), where an offence under this Act has been committed by a Department of Government and it is proved that the offence has been committed with the consent or connivance of, or is attributable to any neglect on the part of, any officer, other than the Head of the Department, such officer shall also be deemed to be guilty of that offence and shall be liable to be proceeded against and punished accordingly.,

42. Protection of Action Taken in Good Faith

No suit, prosecution or other legal proceeding shall lie against the Goverwnent er any officer of the Government or any member or any officer or other employee of the Board in respect of anything which is done or intended to be done in good faith in pursuance of Otis Act or the rules made thereunder.

43. Cognizance of Offences

(1) No court shall take cognizance of any offence under this Act except on a complaint made by-

(a) a Board or any officer authorised in this behalf by it; or

(b) any person who has given notice of not less than sixty days, in the manner prescribed, of the alleged offence and of his intention to make a complaint to the Board or officer authorised as aforesaid, and no court inferior to that of a Metropolitan Magistrate or a Judicial Magistrate of the first class shall try any offence punishable under this Act.

(2) Where a complaint has been made under clause (b) of sub-section (1), the

Board shall, on demand by such person, make available the relevant reports in its possession to that person:

Provided that the Board may refuse to make any such report available to such person if the same is, in its opinion, against the public interest.

44. Members, Officers and Employees of Board to be Public Servants.

All the members and all officers and other employees of a Board when acting or purporting to act in pursuance of any of the provisions of this Act or the rules made thereunder shall be deemed to be public servant within the meaning of section 21 of the Indian Penal Code (45 of 1860).

45. Reports and Returns

The Central Board shall, in relation to its functions under this Act, furnish to the Central Goveniment, and a State Board shall, in relation to its functions under this Act, furnish to the State government and to the Central Board such reports, returns, statistics, accounts and other information as that Government, or, as the case may be, the Central Board may, from time to time, require.

46. Bar of Jurisdiction

No civil court shall have jurisdiction to entertain any suit or proceeding in respect of any matter which an Appellate Authority constituted under this Act is empowered by or under this Act to determine, and no injunction shall be granted by any court or other authority in respect of any action taken or to be taken in pursuance of any power conferred by or under this Act.

CHAPTER VII. MISCELLANEOUS

47. Power of Central Government to Supersede State Board

(1) If at any time the State Government is of opinion-

(a) that a State Board constituted under this Act has persistently made default in the performance of the functions imposed on it by or under this Act, or

(b) that circumstances exist which render it necessary in the public interest so to do, the State Government may, by notification in the Official Gazette, supersede the State Board for such period, not exceeding six months, as may be specified in the notification:

Provided that before issuing a notification under this sub-section for the reasons mentioned in clause (a), the State Government shall give a reasonable opportunity to the State Board to show cause why it should not be superseded and shall consider the explanations and objections, if any, of the State Board.

(2) Upon the publication of a notification under sub-section (1) superseding the State Board,-

(a) all the members shall, as from the date of supersession, vacate their offices as such;

(b) all the powers, functions and duties which may, by or under this Act, be exercised, performed or discharged by the State Board shall, until the State Board is reconstituted under sub-section (3), be exercised, performed or discharged by such person or persons as the State Government may direct.-,

(c) all property owned or controlled by the State Board shall, until the Board is reconstituted under sub-section (3), vest in the State Government.

(3) On the expiration of the period of supersession specified in the notification issued under sub-section (1), the State Government may-

(a) extend the period of supersession for such further term, not exceeding six months, as it may consider necessary; or

(b) reconstitute the State Board by a fresh nomination or appointment as the case may be, and in such case any person who vacated his office under clause (a) of sub-section (2) shall also be eligible for nomination or appointment.

Provided that the State Government may at any time before the expiration of the period of supersession whether originally specified under sub-section (1) or as extended under this sub-section, take action under clause (b) of this sub-section.

48. Special Provision in the Case of Supersession of the Central Board or the State Boards Constituted Under the Water (Prevention and Control of Pollution) Act, 1974.

Where the Central Board or any State Board constituted under the Water (Prevention and Control of Pollution) Act, 1974 (Act 6 of 1974), is superseded by the Central Government or the State Government, as the case may be, under that Act, all the powers, functions and duties of the Central Board or such State Board under this Act shall be exercised, performed or discharged during the period of such supersession by the person or persons, exercising, preforming or discharging the powers, functions and duties of the Central Board or such State Board under the Water (Prevention and Control of Pollution) Act, 1974, during such period.

49. Dissolution of State Boards Constituted Under the Act

(1) As and when the Water (Prevention and Control of Pollution) Act, 1974 (Act 6 of 1974), comes into force in any State and the State Government constitutes a I [Scate Board for the Prevention and Control of Water Pollution] under that Act, the State Board constituted by the State Government under this Act shall stand dissolved and the Board first-mentioned shall exercise the powers and perform the functions of the Board second-mentioned in that State,

(2) On the dissolution of the State Board constituted under this Act,—

(a) all the members shall vacate their offices as such;

(b) all moneys and other property of whatever kind (including the fund of the State Board) owned by, or vested in, the State Board, immediately before such dissolution, shall stand transferred to and vest in the State Board for the Prevention and Control of Water Pollution.

(c) every officer and other employee serving under the State, Board immediately before such dissolution shall be transferred to and become an officer or other employee of the I [State Board for the Prevention and Control of Water Pollution] and hold office by the same tenure and at the same remuneration and on the same terms and conditions of service as he would have held the same if the State Board constituted under this Act had not been dissolved and shall continue to do so unless and until such tenure, remuneration and conditions of service are duly altered by the [State Board for the Prevention and Control of Water Pollution]

Provided that the tenure, remuneration and terms and conditions of service of any such officer or other employee shall not be altered to his disadvantage without the previous sanction of the State Government;

(d) all liabilities obligations of the State Board of whatever kind, immediately before such dissolution, shall be deemed to be the liabilities or obligations, as the case may be, of the l[State Board for the Prevention and Control of Water Pollution] and any proceeding or cause of action, pending or existing immediately before such dissolution by or against the State Board constituted under this Act in relation to such liability or obligation may be continued and enforced by or against the I [State Board for the Prevention and Control of Water Pollution.]

50. [Power to Amend the Schedule.] Rep. by the Air (Prevention and Control of Pollution) Amendment Act, 1987 (47 of 1987), s. 22 (w.e.f. 1-41988).

51. Maintenance of Register

(1) Every State Board shall maintain a register containing particulars of the persons to whom consent has been granted under section 21, the standard for emission laid down by it in relation to each such consent and such other particulars as may be prescribed.

(2) The register maintained under sub-section (1) shall be open to inspection at all reasonable hours by any person interested in or affected by such standards for emission or by any other person authorised by such person in this behalf.

52. Effect of Other Laws

Save as otherwise provided by or under the Atomic Energy Act, 1962 (33 of 1962), in relation to radioactive air pollution the provisions of this Act shall have effect notwithstanding anything inconsistent therewith contained in any enactment other than this Act.

53. Power of Central Government to Make Rules

(1) The Central Government may, in consultation with the Central Board by notification in the Official Gazette, make rules in respect of the following matters namely :-

(a) the intervals and the time and place at which meetings of the Central Board or any committee thereof shall be held and the procedure to be followed at such meetings, including the quorum necessary for the transaction of business thereat, under sub-section (1) of section 10 and under sub-section (2) of section 11;

(b) the fees and allowances to be paid to the members of a committee of the Central Board, not being members of the Board, under sub-section (3) of section 11;

(c) the manner in which and the purposes for which persons may be associated with the Central Board under sub-section (1) of section 12;

(a) the fees and allowance to be paid under sub-section (3) of section 12 to persons associated with the Central Board under sub-section (/) of section 12;

(e) the functions to be performed by the Central Board under clause (j) of sub-section (2) of section 16;

(f) the form in which and the time within which the budget of the Central Board may be prepared and forwarded to the Central Government under section 34;

(i) the form in which the annual report of die Central Board may be prepared under section 35;1

(g) the form in which the accounts of the Central Board may be maintained under sub-section (1) of section 36.

(2) Every rule made by the Central Government under this Act shall be laid, as soon as may be after it is made, before each House of Parliament, while it is in session, for a total period of thirty days which may be comprised in one session or in two or more successive sessions, and if, before the expiry of the session immediately following the session or the successive sessions aforesaid, both Houses agree in making any modification in the rule or both Houses agree that the rule should not be made, the rule shall thereafte have effect only in such modified form or be of no effect, as the case may be; so, however, that any such modification or annulment shall be without prejudice to the validity of anything previously done under that rule.

54. Power of State Government to Make Rules

(1) Subject to the provisions of sub-section (3), the State Government may, by notification in the Official Gazette, make rules to carry out the purposes of this Act in respect of matter not falling within the purview of section 53.

(2) In particular, and without prejudice to the generality of the foregoing power, such rules may provide for all or any of the following matters, namley –

(a) the qualifications, knowledge and experience of scientific, engineering or management aspect of pollution control required for appointment as member-secretary of a State Board constituted under the Act;

(i) the terms and conditions of service of the Chairman and other members (other than the member-secretary) of the State Board constituted under this Act under sub-section (7) of section 7;

(b) the intervals and the time and place at which meetings of the State Board or any committee thereof shall be held and the procedure to be followed at such meetings, including the quorum necessary for the transaction of business thereat, under sub-section (1) of section 10 and under sub-section (2) of section 11;

(c) the fees and allowances to be paid to the members of a committee of the State Board, not being members of the Board under sub-section (3) of section 11;

(d) the manner in which and the purpose for which persons may be associated with the State Board under sub-section (1) of section 12;

(e) the fees and allowances to be paid under sub-section (3) of section 12 to persons associated with the State Board under sub-section (1) of section 12;

(f) the terms and conditions of service of the member-secretary of a State Board constituted under this Act under sub-section (1) of section 14;

(g) the powers and duties to be exercised and discharged by the member-secretary of a State Board under sub-section (2) of section 14;

(h) the conditions subject to which a State Board may appoint such officers and other employees as it considers necessary for the efficient performance of its functions under sub-section (3) of section 14;

(i) the conditions subject to which a State Board may appoint a consultant under sub-section (5) of section 14;

(j) the functions to be performed by the State Board under clause (i) of sub-section (1) of section 17;

(k) the manner in which any area or areas may be declared as air pollution control area or areas under sub-section (1) of section 19;

(l) the form of application for the consent of the State Board, the fees payable therefore, the period within which such application shall be made and the particulars it may contain, under sub-section (2) of section 21;

(m) the procedure to be followed in respect of an inquiry under subsection (3) of section 2 1;

(n) the authorities or agencies to whom information under sub-section (1) of section 23 shall be furnished;

(o) the manner in which samples of air or emission may be taken under sub-section (1) of section 26;

(p) the form of the notice referred to in sub-section (3) of section 26;

(q) the form of the report of the State Board analyst under sub-section (1) of section 27;

(r) the form of the report of the Government analyst under sub-section (3) of section 27;

(s) the functions of the State Air Laboratory, the procedure for the submission to the said Laboratory of samples of air or emission for analysis or tests, the form of Laboratory's report thereon, the fees payable in respect of such report and other matters as may be necessary or expedient to enable that Laboratory to carry out its functions, under sub-section (2) of section 28;

(t) the qualifications required for Government analysts under subsection (1) of section 29;

(u) the qualification required for State Board analysts under sub-section (2) of section 29;

(v) the form and the manner in which appeals may be preferred, the fees payable in respect ot such appeals and the procedure to be followed by the Appellate Authority in disposing of the appeals under sub-section (3) of section 31;

(w) the form in which and the time within which the budget of the State Board may be prepared and forwarded to the State Government under section 34; (ww) the form in which the annual report of the State Board may be prepared under section 35,1;

(x) the form in which the accounts of the State Board may be maintained under the sub-section (1) of section 36;

(i) [the manner in which notice of intention to make a complaint shall be given under section 43;]

(y) the particulars which the register maintained under section 51 may contain;

(z) any other matter which has to be, or may be, prescribed.

(3) After the first constitution of the State Board, no rule with respect to any of the matters referred to in sub-section (2) other than those referred to in clause (aa) thereof, shall be made, varied, amended or repealed without consulting that Board.

Annexure II

Water (Prevention and Control of Pollution) Act, 1974

CHAPTER I : PRELIMINARY

1. Short title, Application and Commencement

(1) This Act may be called the Water (Prevention and Control of Pollution) Act, 1974.

(2) It applies in the first instance to the whole of the States of Assam, Bihar, Gujarat, Haryana, Himachal Pradesh, Jammu and Kashmir, Karnataka, Kerala, Madhya Pradesh, Rajasthan, Tripura and West Bengal and the Union Territories; and it shall apply to such other State which adopts this Act by resolution passed in that behalf under clause (1) of article 252 of the Constitution.

(3) It shall come into force at once in the States of Assam, Bihar, Gujarat, Haryana, Himachal Pradesh, Jammu and Kashmir, Karnataka, Kerala, Madhya Pradesh, Rajasthan, Tripura and West Bengal and in the Union Territories, and in any other State which adopts this Act under clause (1) of article 252 of the Constitution on the date of such adoption and any reference in this Act to the commencement of this

Act shall, in relation to any State or Union Territory, means the date on which this Act comes into force in such State or Union Territory.

2. Definitions

In this Act, unless the context otherwise requires,-

(a) "Board" means the Central Board or a State Board;

[(b) "Central Board" means the Central Pollution Control Board constituted under section 3;

(c) "member" means a member of a Board and includes the Chairman thereof;

[(d) "occupier" in relation to any factory or premises, means the person who has control over the affairs of the factory or the premises, and includes, in relation to any substance, the person in possession of the substance;

[(i) "outlet" includes any conduit pipe or channel, open or closed, carrying sewage or trade effluent or any other holding arrangement which causes or is likely to cause, pollution;

(e) "pollution" means such contamination of water or such alteration of the physical, chemical or biological properties of water or such discharge of any sewage or trade effluent or of any other liquid, gaseous or solid substance into water (whether directly or indirectly) as may, or is likely to, create a nuisance or render such water harmful or injurious to public health or safety, or to domestic, commercial, industrial, agricultural or other legitimate uses, or to the life and health of animals or plants or of aquatic organisms;

(f) "prescribed" means prescribed by rules made under this Act by the Central Government or, as the case may be, the State Government;

(g) "sewage effluent" means effluent from any sewerage system or sewage disposal works and includes sullage from open drains;

(i) "State Government" in relation to a Union Territory means the Administrator thereof appointed under article 239 of the Constitution;

(j) "stream" includes-

(i) river;

(ii) water course (whether flowing or for the time being dry);

(iii) inland water (whether natural or artificial);

(iv) sub-terranean waters;

(v) sea or tidal waters to such extent or, as the case may be, to such point as the State Government may, by notification in the Official Gazette, specify in this behalf;

(k) "trade effluent" includes any liquid, gaseous or solid substance which is discharged from any premises used for carrying on any [industry, operation or process, or treatment and disposal system], other than domestic sewage.

CHAPTER II : THE CENTRALAND STATE BOARDS FOR PREVENTION AND CONTROL OF WATER POLLUTION

3. Constitution of Central Board

(1) The Central Government shall, with effect from such date (being a date not later than six months of the commencement of this Act in the States of Assam, Bihar, Gujarat, Haryana, Himachal Pradesh, Jammu

and Kashmir, Karnataka, Kerala, Madhya Pradesh, Rajasthan, Tripura and West Bengal and in the Union Territories) as it may, by notification in the Official Gazette, appoint, constitute a Central Board to be called the [Central Pollution Control Board] to exercise the powers conferred on and perform the functions assigned to that Board under this Act.

(2) The Central Board shall consist of the following members, namely,-

(a) a full-time Chairman, being a person having special knowledge or practical experience in respect of [matters relating to environmental protection] or a person having knowledge and experience in administering institutions dealing with the matters aforesaid, to be nominated by the Central Government;

(b) [such number of officials, not exceeding five], to be nominated by the Central Government to represent that government;

(c) such number of persons, not exceeding five to be nominated by the Central Government, from amongst the members of the State Boards, of whom not exceeding two shall be from those referred to in clause (c) of sub-section (2) of section 4;

(d) such number of non-officials, not exceeding three, to be nominated by the Central Government, to represent the interests of agriculture, fishery or industry or trade or any other interest which, in the opinion of the Central Government, ought to be represented;

(e) two persons to represent the companies or corporations owned, controlled or managed by the Central Government, to be nominated by that government;

(f) a full-time member-secretary, possessing qualifications, knowledge and experience of scientific, engineering or management aspects of pollution control, to be appointed by the Central Government.

(3) The Central Board shall be a body corporate with the name aforesaid having perpetual succession and a common seal with power, subject to the provisions of this Act, to acquire, hold and dispose of property and to contract, and may, by the aforesaid name, sue or be sued.

4. Constitution of State Boards

(1) The State Government shall, with effect from such date as it may, by notification in the Official Gazette, appoint, constitute a [State Pollution Control Board] under such name as may be specified in the notification, to exercise the powers conferred on and perform the functions assigned to that Board under this Act.

(2) A State Board shall consist of the following members, namely,-

(a) a Chairman, being, a person having special knowledge or practical experience in respect of 5[matters relating to environmental protection] or a person having knowledge and experience in administering institutions dealing with the matters aforesaid, to be nominated by the State Government:

PROVIDED that the Chairman may be either whole-time or part-time as the State Government may think fit;

(b) such number of officials, not exceeding five, to be nominated by the State Government to represent that government;

(c) such number of persons, not exceeding five, to be nominated by the State Government from amongst the members of the local authorities functioning within the State;

(d) such number of non-officials, not exceeding three to be nominated by the State Government to represent the interests of agriculture, fishery or industry or trade or any other interest which, in the opinion of the State Government, ought to be represented;

(e) two persons to represent the companies or corporations owned, controlled or managed by the State Government, to be nominated by that government;

(f) a full-time member-secretary, possessing qualifications, knowledge and experience of scientific, engineering or management aspects of pollution control, to be appointed by the State Government.

(3) Every State Board shall be a body corporate with the name specified by the State Government in the notification under sub-section (1), having perpetual succession and a common seal with power, subject to the provisions of this Act, to acquire hold and dispose of property and to contract, and may, by the said name, sue or be sued.

(4) Notwithstanding anything contained in this section, no State Board shall be constituted for a Union Territory and in relation to a Union Territory, the Central Board shall exercise the powers and perform the functions of a State Board for that Union Territory:

PROVIDED that in relation to any Union Territory the Central Board may delegate all or any of its powers and functions under this sub-section to such person or body of persons as the Central Government may specify.

5. Terms and Conditions of Service of Members

(1) Save as otherwise provided by or under this Act, a member of a Board, other than a member-secretary, shall hold office for a term of three years from the date of his nomination:

PROVIDED that a member shall, notwithstanding the expiration of his term, continue to hold office until his successor enters upon his office.

(2) The term of office of a member of a Board nominated under clause (b) or clause (e) of sub-section (2) of section 3 or clause (b) or clause (e) of sub-section (2) of section 4 shall come to an end as soon as he ceases to hold the office under the Central Government or the State Government or, as the case may be, the company or corporation owned, controlled or managed by the Central Government or the State Government, by virtue of which he was nominated.

(3) The Central Government or, as the case may be, the State Government may, if it thinks fit, remove any member of a Board before the expiry of his term of office, after giving him a reasonable opportunity of showing cause against the same.

(4) A member of a Board, other than the member-secretary, may at any time resign his office by writing under his hand addressed-

(a) in the case of the Chairman, to the Central Government or, as the case may be, the State Government; and

(b) in any other case, to the Chairman of the Board; and the seat of the Chairman or such other member shall thereupon become vacant.

(5) A member of a Board, other than the member-secretary, shall be deemed to have vacated his seat if he is absent without reason, sufficient in the opinion of the Board, from three consecutive meetings of the Board, [or where he is nominated under clause (c) or clause (e) of sub-section (2) of section 3 or under clause (c) or clause (e) of sub-section (2) of section 4, if he ceases to be a member of the State Board or of the local authority or, as the case may be, of the company or corporation owned, controlled or managed by the Central Government or the State Government and such vacation of seat shall, in either case, take effect from such date as the Central Government or, as the case may be, the State Government may, by notification in the Official Gazette, specify].

(6) A casual vacancy in a Board shall be filled by a fresh nomination and the person nominated to fill the vacancy shall hold office only for the remainder of the term for which the member in whose place he was nominated.

(7) A member of a Board [shall be eligible for renomination].

(8) The other terms and conditions of service of a member of a Board, other than the Chairman and member-secretary, shall be such as may be prescribed.

(9) The other terms and conditions of service of the Chairman shall be such as may be prescribed.

6. Disqualifications

(1) No person shall be a member of Board, who-

(a) is, or at any time has been adjudged insolvent or has suspended payment of his debts or has compounded with his creditors, or

(b) is of unsound mind and stands so declared by a competent court, or

(c) is, or has been, convicted of an offence which, in the opinion of the Central Government or, as the case may be, of the State Government, involves moral turpitude, or

(d) is, or at any time has been, convicted of an offence under this Act, or

(e) has directly or indirectly by himself or by any partner, any share or interest in any firm or company carrying on the business of manufacture, sale or hire of machinery, plant, equipment, apparatus or fittings for the treatment of sewage or trade effluents, or

(f) is a director or a secretary, manager or other salaried officer or employee of any company or firm having any contract with the Board, or with the government constituting the Board, or with a local authority in the State, or with a company or corporation owned, controlled or managed by the government, for the carrying out of sewerage schemes or for the installation of plants for the treatment of sewage or trade effluents, or

(g) has so abused, in the opinion of the Central Government or as the case may be, of the State Government, his position as a member, as to render his continuance on the Board detrimental to the interest of the general public.

(2) No order of removal shall be made by the Central Government or the State Government, as the case may be, under this section unless the member concerned has been given a reasonable opportunity of showing cause against the same.

(3) Notwithstanding anything contained in sub-sections (1) and (7) of section 5, a member who has been

removed under this section shall not be eligible for renomination as a member.

7. Vacation of Seats by Members

If a member of a Board becomes subject to any of the disqualifications specified in section 6, his seat shall become vacant.

8. Meetings of Boards

A Board shall meet at least once in every three months and shall observe such rules of procedure in regard to the transaction of business at its meetings as may be prescribed:

PROVIDED that if, in the opinion of the Chairman, any business of an urgent nature is to be transacted, he may convene a meeting of the Board at such time as he thinks fit for the aforesaid purpose.

9. Constitution of Committees

(1) A Board may constitute as many committees consisting wholly of members or wholly of other persons or partly of members and partly of other persons, and for such purpose or purposes as it may think fit.

(2) A committee constituted under this section shall meet at such time and at such place, and shall observe such rules of procedure in regard to the transaction of business at its meetings, as may be prescribed.

(3) The members of a committee (other than the members of the Board) shall be paid such fees and allowances, for attending its meetings and for attending to any other work of the Board as may be prescribed.

10. Temporary Association of Persons with Board for Particular Purposes

(1) A Board may associate with itself in such manner, and for such purposes, as may be prescribed any

person whose assistance or advice it may desire to obtain in performing any of its functions under this Act.

(2) A person associated with the Board under sub-section (1) for any purpose shall have a right to take part in the discussions of the Board relevant to that purpose, but shall not have a right to vote at a meeting of the Board, and shall not be a member for any other purpose.

(3) A person associated with the Board under sub-section (1) for any purpose shall be paid such fees and allowances, for attending its meetings and for attending to any other work of the Board, as may be prescribed.

11. Vacancy in Board not to Invalidate Acts or Proceedings

No act or proceeding of a Board or any committee thereof shall be called in question on the ground merely of the existence of any vacancy in, or any defect in the constitution of, the Board or such committee, as the case may be.

11A. Delegation of Powers to Chairman

The Chairman of a Board shall exercise such powers and perform such duties as may be prescribed or as may, from time to time, be delegated to him by the Board.

12. Member-Secretary and Officers and Other Employees of Board

(1) The terms and conditions of service of the member-secretary shall be such as may be prescribed.

(2) The member-secretary shall exercise such powers and perform such duties as may be prescribed or as may,

from time to time, be delegated to him by the Board or its Chairman.

(3) Subject to such rules as may be made by the Central Government or, as the case may be, the State Government in this behalf, a Board may appoint such officers and employees as it considers necessary for the efficient performance of its functions.

(3i) The method of recruitment and the terms and conditions of service (including the scales of pay) of the officers (other than the member-secretary) and other employees of the Central Board or a State Board shall be such as may be determined by regulations made by the Central Board or, as the case may be, by the State Board:

PROVIDED that no regulation made under this sub-section shall take effect unless-

(a) in the case of a regulation made by the Central Board, it is approved by the Central Government; and

(b) in the case of a regulation made by a State Board, it is approved by the State Government.

(3ii) The Board may, by general or special order, and subject to such conditions and limitations, if any, as may be specified in the order, delegate to any officer of the Board such of its powers and functions under this Act as it may deem necessary.

(4) Subject to such conditions as may be prescribed, a Board may from time to time appoint any qualified person to be a consulting engineer to the Board and pay him such salaries and allowances and subject him to such other terms and conditions of service as it thinks fit.

CHAPTER III : JOINT BOARDS

13. Constitution of Joint Boards

(1) Notwithstanding anything contained in this Act, an agreement may be entered into-

(a) by two or more governments of contiguous States, or

(b) by the Central Government (in respect of one or more Union Territories) and one or more governments of States contiguous to such Union Territory or Union Territories, to be in force for such period and to be subject to renewal for such further period, if any, as may be specified in the agreement to provide for the constitution of a Joint Board-

(i) in a case referred to in clause (a), for all the participating States, and

(ii) in a case referred to in clause (b), for the participating Union Territory or Union Territories and the State or States.

(2) An agreement under this section may-

(a) provide, in a case referred to in clause (a) of sub-section (1), for the apportionment between the participating States and in a case referred to in clause (b) of that sub-section, for the apportionments between the Central Government and the participating State Government or State Governments, of the expenditure in connection with the Joint Board;

(b) determine, in a case referred to in clause (a) of sub-section (1), which of the participating State Governments and in a case referred to in clause (b) of that sub-section, whether the Central Government or the participating State Government (if there are more than one participating State, also which of the participating State Governments) shall exercise and perform the several powers and functions of the State Government under this Act and the references in this Act to the State Government shall be construed accordingly;

(c) provide for consultation, in a case referred to in clause (a) of sub-section (1), between the participating State Governments and in a case referred to in clause (b) of that sub-section, between the Central Government and the participating State Government or State Governments either generally or with reference to particular matters arising under this Act;

(d) make such incidental and ancillary provisions, not inconsistent with this Act, as may be deemed necessary or expedient for giving effect to the agreement.

(3) An agreement under this section shall be published, in a case referred to in clause (a) of sub-section (1), in the Official Gazette of the participating States and in a case referred to in clause (b) of that sub-section, in the Official Gazette of the participating Union Territory or Union Territories and participating State or States.

14. Composition of Joint Boards

(1) A Joint Board constituted in pursuance of an agreement entered into under clause (a) of sub-section (1) of section 13 shall consist of the following members namely,-

(a) a full-time chairman, being a person having special knowledge or practical experience in respect of 5[matters relating to environmental protection] or a person having knowledge and experience in administering institutions dealing with the matters aforesaid, to be nominated by the Central Government;

(b) two officials from each of the participating States to be nominated by the concerned participating State Government to represent that government;

(c) one person to be nominated by each of the participating State Governments from amongst the members of the local authorities functioning within the State concerned;

(d) one non-official to be nominated by each of the participating State Governments to represent the interests of agriculture, fishery or industry or trade in the State concerned or any other interest which, in the opinion of the participating State Government, is to be represented;

(e) two persons to be nominated by the Central Government to represent the companies or corporations owned, controlled or managed by the participating State Governments;

(f) a full-time member-secretary, possessing qualifications, knowledge and experience of scientific, engineering or management aspects of pollution control, to be appointed by the Central Government.

(2) A Joint Board constituted in pursuance of an agreement entered into under clause (b) of sub-section (1) of section 13 shall consist of the following members, namely,-

(a) a full-time Chairman, being a person having special knowledge or practical experience in respect of 5[matters relating to environmental protection] or a person having knowledge and experience in administering institutions dealing with the matters aforesaid, to be nominated by the Central Government;

(b) two officials to be nominated by the Central Government from the participating Union Territory or each of the participating Union Territories, as the case may be, and two officials to be nominated, from the participating State or each of the participating States, as the case may be, by the concerned participating State Government;

(c) one person to be nominated by the Central Government from amongst the members of the local authorities functioning within the participating Union Territory or each of the participating Union Territories, as the case may be, and one person to be nominated, from amongst the members of the local authorities functioning within the participating State or each of the participating States, as the case may be, by the concerned participating State Government;

(d) one non-official to be nominated by Central Government and one person to be nominated by the participating State Government or State Governments to represent the interests of agriculture, fishery or industry or trade in the Union Territory or in each of the Union Territories or the State or in each of the States, as the case may be, or any other interest which in the opinion of the Central Government or, as the case may be, of the State Government is to be represented;

(e) two persons to be nominated by the Central Government to represent the companies or corporations owned, controlled or managed by the

Central Government and situate in the participating Union Territory or Territories and two persons to be nominated by the Central Government to represent the companies or corporations owned, controlled or managed by the participating State Governments;

(f) a full-time member-secretary, possessing qualifications, knowledge and experience of scientific, engineering or management aspects of pollution control, to be appointed by the Central Government.

(3) When a Joint Board is constituted in pursuance of an agreement under clause (b) of sub-section (1) of section 13, the provisions of sub-section (4) of section 4 shall cease to apply in relation to the Union Territory for which the Joint Board is constituted.

(4) Subject to the provisions of sub-section (3), the provisions of sub-section (3) of section 4 and sections 5 to 12 (inclusive) shall apply in relation to the Joint Board and its member-secretary as they apply in relation to a State Board and its member-secretary.

(5) Any reference in this Act to the State Board shall, unless the context otherwise requires, be construed as including a Joint Board.

15. Special Provisions Relating to Giving of Directions

Notwithstanding anything contained in this Act where any Joint Board is constituted under section 13-

(a) the government of the State for which the Joint Board is constituted shall be competent to give any direction under this Act only in cases where such direction relates to a matter within the exclusive territorial jurisdiction of the State;

(b) the Central Government alone shall be competent to give any direction under this Act where such direction relates to a matter within the territorial jurisdiction of two or more States or pertaining to a Union Territory.

CHAPTER IV : POWERS AND FUNCTIONS OF BOARDS

16. Functions of Central Board

(1) Subject to the provisions of this Act, the main function of the Central Board shall be to promote cleanliness of streams and wells in different areas of the States.

(2) In particular and without prejudice to the generality of the foregoing function, the Central Board may perform all or any of the following functions, namely,-

(a) advise the Central Government on any matter concerning the prevention and control of water pollution;

(b) co-ordinate the activities of the State Boards and resolve disputes among them;

(c) provide technical assistance and guidance to the State Boards, carry out and sponsor investigations and research relating to problems of water pollution and prevention, control or abatement of water pollution;

(d) plan and organise the training of persons engaged or to be engaged in programmes for the prevention, control or abatement of water pollution on such terms and conditions as the Central Board may specify;

(e) organise through mass media a comprehensive programme regarding the prevention and control of water pollution;

(i) perform such of the functions of any State Board as may be specified in an order made under sub-section (2) of section 18;

(f) collect, compile and publish technical and statistical data relating to water pollution and the measures devised for its effective prevention and control and prepare manuals, codes or guides relating to treatment and disposal of sewage and trade effluents and disseminate information connected therewith;

(g) lay down, modify or annul, in consultation with the State Government concerned, the standards for a stream or well:

PROVIDED that different standards may be laid down for the same stream or well or for different streams or wells, having regard to the quality of water, flow characteristics of the stream or well and the nature of the use of the water in such stream or well or streams or wells;

(h) plan and cause to be executed a nation-wide programme for the prevention, control or abatement of water pollution;

(i) perform such other functions as may be prescribed.

(3) The Board may establish or recognise a laboratory or laboratories to enable the Board to perform its functions under this section efficiently, including the analysis of samples of water from any stream or well or of samples of any sewage or trade effluents.

17. Functions of State Board

(1) Subject to the provisions of this Act, the functions of a State Board shall be-

(a) to plan a comprehensive programme for the prevention, control or abatement of pollution of streams and wells in the State and to secure the execution thereof;

(b) to advise the State Government on any matter concerning the prevention, control or abatement of water pollution;

(c) to collect and disseminate information relating to water pollution and the prevention, control or abatement thereof;

(d) to encourage, conduct and participate in investigations and research relating to problems of water pollution and prevention, control or abatement of water pollution;

(e) to collaborate with the Central Board in organising the training of persons engaged or to be engaged in programmes relating to prevention, control or abatement of water pollution and to organise mass education programmes relating thereto;

(f) to inspect sewage or trade effluents, works and plants for the treatment of sewage and trade effluents and to review plans, specifications or other data relating to plants set up for the treatment of water, works for the purification thereof and the system for the disposal of sewage or trade effluents or in connection with the grant of any consent as required by this Act;

(g) to lay down, modify or annul effluent standards for the sewage and trade effluents and for the quality of receiving waters (not being water in an inter-State stream) resulting from the discharge of effluents and to classify waters of the State;

(h) to evolve economical and reliable methods of treatment of sewage and trade effluents, having regard to the peculiar conditions of soils, climate and water resources of different regions and more especially the prevailing flow characteristics of water in streams and wells which render it impossible to attain even the minimum degree of dilution;

(i) to evolve methods of utilisation of sewage and suitable trade effluents in agriculture;

(j) to evolve efficient methods of disposal of sewage and trade effluents on land, as are necessary on account of the predominant conditions of scant stream flows that do not provide for major part of the year the minimum degree of dilution;

(k) to lay down standards of treatment of sewage and trade effluents to be discharged into any particular stream taking into account the minimum fair weather dilution available in that stream and the tolerance limits of pollution permissible in the water of the stream, after the discharge of such effluents;

(l) to make, vary or revoke any order-

(i) for the prevention, control or abatement of discharges of waste into streams or wells;

(ii) requiring any person concerned to construct new systems for the disposal of sewage and trade effluents or to modify, alter or extend any such existing system or to adopt such remedial measures as are necessary to prevent, control or abate water pollution;

(m) to lay down effluent standards to be complied with by persons while causing discharge of sewage or sullage or both and to lay down, modify or annul effluent standards for the sewage and trade effluents;

(n) to advise the State Government with respect to the location of any industry the carrying on of which is likely to pollute a stream or well;

(o) to perform such other functions as may be prescribed or as may, from time to time, be entrusted to it by the Central Board or the State Government.

(2) The Board may establish or recognise a laboratory or laboratories to enable the Board to perform its functions under this section efficiently, including the analysis of samples of water from any stream or well or of samples of any sewage or trade effluents.

18. Power to Give Directions

(1) [In the performance of its functions under this Act-]

(a) the Central Board shall be bound by such directions in writing as the Central Government may give to it; and

(b) every State Board shall be bound by such directions in writing as the Central Board or the State Government may give to it:

PROVIDED that where a direction given by the State Government is inconsistent with the direction given by the Central Board, the matter shall be referred to the Central Government for its decision.

(2) Where the Central Government is of the opinion that any State Board has defaulted in complying with any directions given by the Central Board under sub-section (1) and as a result of such default a grave emergency has arisen and it is necessary or expedient so to do in the public interest, it may, by order, direct the Central Board to perform any of the functions of the State Board in relation to such area for such period and for such purposes, as may be specified in the order.

(3) Where the Central Board performs any of the functions of the State Board in pursuance of a direction under sub-section (2), the expenses, if any, incurred by the Central Board with respect to the performance of such functions may, if the State Board is empowered to recover such expenses, be recovered by the Central Board with interest (at such reasonable rate as the Central Government may, by order, fix) from the date when a demand for such expenses is made until it is paid from the person or persons concerned as arrears of land revenue or of public demand.

(4) For the removal of doubts, it is hereby declared that any directions to perform the functions of any State Board given under sub-section (2) in respect of any area would not preclude the State Board from performing such functions in any other area in the State or any of its other functions in that area.]

CHAPTER V :PREVENTION AND CONTROL OF WATER POLLUTION

19. Power of State Government to Restrict the Application of the Act to Certain Areas

(1) Notwithstanding anything contained in this Act, if the State Government, after consultation with, or on the recommendation of the State Board, is of opinion that the provisions of this Act need not apply to entire State, it may, by notification in the Official Gazette, restrict the application of this Act to such area or areas as may be declared therein as water pollution, prevention and control area or areas and thereupon the provisions of this Act shall apply only to such area or areas.

(2) Each water pollution, prevention and control area may be declared either by reference to a map or by reference to the line of any watershed or the boundary of any district or partly by one method and partly by another.

(3) The State Government may, by notification in the Official Gazette-

(a) alter any water pollution, prevention and control area whether by way of extension or reduction; or

(b) define a new water pollution, prevention and control area in which may be merged one or more water pollution, prevention and control areas, or any part or parts thereof.

20. Power to Obtain Information

(1) For the purpose of enabling a State Board to perform the functions conferred on it by or under this Act, the State Board or any officer empowered by it in that behalf, may make surveys of any area and gauge and keep records of the flow or volume and other characteristics of any stream or well in such area, and may take steps for the measurement and recording of the rainfall in such area or any part thereof and for the installation and maintenance for those purposes of gauges or other apparatus and works connected therewith, and carry out stream surveys and may take such other steps as may be necessary in order to obtain any information required for the purposes aforesaid.

(2) A State Board may give directions requiring any person who in its opinion is abstracting water from any such stream or well in the area in quantities which are substantial in relation to the flow or volume of

that stream or well or is discharging sewage or trade effluent into any such stream or well, give such information as to the abstraction or the discharge at such times and in such form as may be specified in the directions.

(3) Without prejudice to the provisions of sub-section (2), a State Board may, with a view to preventing or controlling pollution of water, give directions requiring any person in charge of any establishment where any 17[industry, operation or process, or treatment and disposal system] is carried on, to furnish to it information regarding the construction, installation or operation of such establishment or of any disposal system or of any extension or addition thereto in such establishment and such other particulars as may be prescribed.

21. Power to Take Samples of Effluents and Procedure to be Followed in Connection Therewith

(1) A State Board or any officer empowered by it in this behalf shall have power to take for the purpose of analysis samples of water from any stream or well or samples of any sewage or trade effluent which is passing from any plant or vessel or from or over any place into any such stream or well.

(2) The result of any analysis of a sample of any sewage or trade effluent taken under sub-section (1) shall not be admissible in evidence in any legal proceeding unless the provisions of sub-sections (3), (4) and (5) are complied with.

(3) Subject to the provisions of sub-sections (4) and (5), when a sample (composite or otherwise as may be warranted by the process used) of any sewage or trade effluent is taken for analysis under sub-section

(1), the person taking the sample shall-

(a) serve on the person in charge of, or having control over, the plant or vessel or in occupation of the place (which person is hereinafter referred to as the occupier) or any agent of such occupier, a notice, then and there in such form as may be prescribed of his intention to have it so analysed;

(b) in the presence of the occupier or his agent, divide the sample into two parts;

(c) cause each Part to be placed in a container which shall be marked and sealed and shall also be signed both by the person taking the sample and the occupier or his agent;

(d) send one container forthwith-

(i) in a case where such sample is taken from any area situated in a Union Territory, to the laboratory established or recognised by the Central Board under section 16; and

(ii) in any other case, to the laboratory established or recognised by the State Board under section 17;

(e) on the request of the occupier or his agent, send the second container-

(i) in a case where such sample is taken from any area situated in a Union Territory, to the laboratory established or specified under sub-section (1) of section 51; and

(ii) in any other case, to the laboratory established or specified under sub-section (1) of section 52.

(4) [When a sample of any sewage or trade affluent is taken for analysis under sub-section (1) and the person taking the sample serves on the occupier or his agent, a notice under clause (a) of sub-section (3) and the occupier or his agent wilfully absents himself, then-

(a) the sample so taken shall be placed in a container which shall be marked and sealed and shall also be signed by the person taking the sample and the same shall be sent forthwith by such person for analysis to the laboratory referred to in sub-clause (i) or sub-clause (ii), as the case may be, of clause (e) of sub-section (3) and such person shall inform the government analyst appointed under sub-section (1) or sub-section (2), as the case may be, of section 53, in writing about the wilful absence of the occupier or his agent; and

(b) the cost incurred in getting such sample analysed shall be payable by the occupier or his agent and in case of default of such payment, the same shall be recoverable from the occupier or his agent, as the case may be, as an arrear of land revenue or of public demand:

PROVIDED that no such recovery shall be made unless the occupier or, as the case may be, his agent has been given a reasonable opportunity of being heard in the matter.

(5) When a sample of any sewage or trade effluent is taken for analysis under sub-section (1) and the person taking the sample serves on the occupier or his agent a notice under clause (a) of sub-section (3) and the occupier or his agent who is present at the time of taking the sample does not make a request for dividing the sample into two parts as provided in clause (b) of sub-section (3), then, the sample so taken shall be placed in a container which shall be marked and sealed and shall also be signed by the person taking the sample and the same shall be sent forthwith by such person for analysis to the laboratory referred to in sub-clause (i) or sub-clause

(ii), as the case may be, of clause (d) of sub-section (3).

22. Reports of the Result of Analysis on Samples Taken Under Section 21

(1) Where a sample of any sewage or trade effluent has been sent for analysis to the laboratory established or recognised by the Central Board or, as the case may be, the State Board, the concerned Board analyst appointed under sub-section (3) of section 53 shall analyse the sample and submit a report in the prescribed form of the result of such analysis in triplicate to the Central Board or the State Board as the case may be.

(2) On receipt of the report under sub-section (1), one copy of the report shall be sent by the Central Board or the State Board, as the case may be, to the occupier or his agent referred to in section 21, another copy shall be preserved for production before the court in case any legal proceedings are taken against him and the other copy shall be kept by the concerned Board.

(3) Where a sample has been sent for analysis under clause (e) of sub-section (3) or sub-section (4) of section 21 to any laboratory mentioned therein, the government analyst referred to in that sub-section shall analyse the sample and submit a report in the prescribed form of the result of the analysis in triplicate to the Central Board or, as the case may be, the State Board which shall comply with the provisions of sub-section (2).

(4) If there is any inconsistency or discrepancy between, or variation in the results of, the analysis carried out by the laboratory established or recognised by the Central Board or the State Board, as the case may

be, and that of the laboratory established or specified under section 51 or section 52, as the case may be, the report of the latter shall prevail.

(5) Any cost incurred in getting any sample analysed at the request of the occupier or his agent shall be payable by such occupier or his agent and in case of default the same shall be recoverable from him as arrears of land revenue or of public demand.

23. Power of Entry and Inspection

(1) Subject to the provisions of this section, any person empowered by a State Board in this behalf shall have a right at any time to enter, with such assistance as he considers necessary, any place-

(a) for the purpose of performing any of the functions of the Board entrusted to him;

(b) for the purpose of determining whether and if so in what manner, any such functions are to be performed or whether any provisions of this Act or the rules made thereunder of any notice, order, direction or authorisation served, made, given, or granted under this Act is being or has been complied with;

(c) for the purpose of examining any plant, record, register, document or any other material object or for conducting a search of any place in which he has reason to believe that an offence under this Act or the rules made thereunder has been or is being or is about to be committed and for seizing any such plant, record, register, document or other material object, if he has reason to believe that it may furnish evidence of the commission of an offence punishable under this Act or the rules made thereunder:

PROVIDED that the right to enter under this sub-section for the inspection of a well shall be exercised only at reasonable hours in a case where such well is situated in any premises used for residential purposes and the water thereof is used exclusively for domestic purposes.

(2) The provisions of the 19[Code of Criminal Procedure, 1973 (2 of 1974)], or, in relation to the State of Jammu and Kashmir, the provisions of any corresponding law in force in that State, shall, so far as may be, apply to any search or seizure under this section as they apply to any search or seizure made under the authority of a warrant issued under [section 94] of the said Code, or, as the case may be, under the corresponding provisions of the said law.

Explanation : For the purposes of this section, "place" includes vessel.

24. Prohibition on Use of Stream or Well for Disposal of Polluting Matter, etc.

(1) Subject to the provisions of this section-

(a) no person shall knowingly cause or permit any poisonous, noxious or polluting matter determined in accordance with such standards as may be laid down by the State Board to enter (whether directly or indirectly) into any [Stream or well or sewer or on land]; or

(b) no person shall knowingly cause or permit to enter into any stream any other matter which may tend, either directly or in combination with similar matters, to impede the proper flow of the water of the stream in a manner leading or likely to lead to a substantial aggravation of pollution due to other causes or of its consequences.

(2) A person shall not be guilty of an offence under sub-section (1), by reason only of having done or caused to be done any of the following acts, namely,-

(a) constructing, improving or maintaining in or across or on the bank or bed of any stream any building, bridge, weir, dam, sluice, dock, pier, drain or sewer or other permanent works which he has a right to construct, improve or maintain;

(b) depositing any materials on the bank or in the bed of any stream for the purpose of reclaiming land, or for supporting, repairing or protecting the bank or bed of such stream provided such materials are not capable of polluting such stream;

(c) putting into any stream any sand or gravel or other natural deposit which has flowed from or been deposited by the current of such stream;

(d) causing or permitting, with the consent of the State Board, the deposit accumulated in a well, pond or reservoir to enter into any stream.

(3) The State Government may, after consultation with, or on the recommendation of, the State Board, exempt, by notification in the Official Gazette, any person from the operation of sub-section (1) subject to such conditions, if any, as may be specified in the notification and any condition so specified may by a like notification be altered, varied or amended.

25. Restrictions on New Outlets and New Discharges

(1) Subject to the provisions of this section, no person shall, without the previous consent of the State Board-

(a) establish or take any steps to establish any industry, operation or process, or any treatment and disposal system or any extension or addition thereto, which is likely to discharge sewage or trade effluent into a stream or well or sewer or on land (such discharge being hereafter in this section referred to as discharge of sewage); or

(b) bring into use any new or altered outlet for the discharge of sewage; or

(c) begin to make any new discharge of sewage:

PROVIDED that a person in the process of taking any steps to establish any industry, operation or process immediately before the commencement of the Water (Prevention and Control of Pollution) Amendment Act, 1988, for which no consent was necessary prior to such commencement, may continue to do so for a period of three months from such commencement or, if he has made an application for such consent, within the said period of three months, till the disposal of such application.

(2) An application for consent of the State Board under sub-section (1) shall be made in such form, contain such particulars and shall be accompanied by such fees as may be prescribed.

(3) The State Board may make such inquiry as it may deem fit in respect of the application for consent referred to in sub-section (1) and in making any such inquiry shall follow such procedure as may be prescribed.

(4) The State Board may-

(a) grant its consent referred to in sub-section (1), subject to such conditions as it may impose, being-

(i) in cases referred to in clauses (a) and (b) of sub-section (1) of section 25, conditions as to the point of discharge of sewage or as to the use of that outlet or any other outlet for discharge of sewage;

(ii) in the case of a new discharge, conditions as to the nature and composition, temperature, volume or rate of discharge of the effluent from the land or premises from which the discharge or new discharge is to be made; and

(iii) that the consent will be valid only for such period as may be specified in the order, and any such conditions imposed shall be binding on any person establishing or taking any steps to establish any industry, operation or process, or treatment and disposal system or extension or addition thereto, or using the new or altered outlet, or discharging the effluent from the land or premises aforesaid; or

(b) refuse such consent for reasons to be recorded in writing.

(5) Where, without the consent of the State Board, any industry, operation or process, or any treatment and disposal system or any extension or addition thereto, is established, or any steps for such establishment have been taken or a new or altered outlet is brought into use for the discharge of sewage or a new discharge of sewage is made, the State Board may serve on the person who has established or taken steps to establish any industry, operation or process, or any treatment and disposal system or any extension or addition thereto, or using the outlet, or making the discharge, as the case may be, a notice imposing any such conditions as it might have imposed on an application for its consent in respect of such establishment, such outlet or discharge.

(6) Every State Board shall maintain a register containing particulars of the conditions imposed under this section and so much of the register as relates to any outlet, or to any effluent, from any land or premises shall be open to inspection at all reasonable hours by

any person interested in, or affected by such outlet, land or premises, as the case may be, or by any person authorised by him in this behalf and the conditions so contained in such register shall be conclusive proof that the consent was granted subject to such conditions.

(7) The consent referred to in sub-section (1) shall, unless given or refused earlier, be deemed to have been given unconditionally on the expiry of a period of four months of the making of an application in this behalf complete in all respects to the State Board.

(8) For the purposes of this section and sections 27 and 30-

(a) the expression "new or altered outlet" means any outlet which is wholly or partly constructed on or after the commencement of this Act or which (whether so constructed or not) is substantially altered after such commencement;

(b) the expression "new discharge" means a discharge which is not, as respects the nature and composition, temperature, volume, and rate of discharge of the effluent substantially a continuation of a discharge made within the preceding twelve months (whether by the same or a different outlet), so however that a discharge which is in other respects a continuation of previous discharge made as aforesaid shall not be deemed to be a new discharge by reason of any reduction of the temperature or volume or rate of discharge of the effluent as compared with the previous discharge.

26. Provision Regarding Existing Discharge of Sewage or Trade Effluent

Where immediately before the commencement of this Act any person was discharging any sewage or trade effluent into a [stream or well or sewer or on land][21], the provisions of section 25 shall, so far as may be, apply in relation to such person as they apply in relation to the person referred to in that section subject to the modification that the application for consent to be made under sub-section (2) of that section [shall be made on or before such date as may be specified by the State Government by notification in this behalf in the Official Gazette.

27. Refusal or Withdrawal of Consent by State Board

(1) A State Board shall not grant its consent under sub-section (4) of section 25 for the establishment of any industry, operation or process, or treatment and disposal system or extension or addition thereto, or to the bringing into use of a new or altered outlet unless the industry, operation or process, or treatment and disposal system or extension or addition thereto, or the outlet is so established as to comply with any conditions imposed by the Board to enable it to exercise its right to take samples of the effluent.

(2) A State Board may from time to time review-

(a) any condition imposed under section 25 or section 26 and may serve on the person to whom a consent under section 25 or section 26 is granted a notice making any reasonable variation of or revoking any such condition.

(b) the refusal of any consent referred to in sub-section (1) of section 25 or section 26 or the grant of such consent without any condition, and may make such orders as it deemed fit.

(3) Any condition imposed under section 25 or section 26 shall be subject to any variation made under sub-

section (2) and shall continue in force until revoked under that sub-section.

28. Appeals

(1) Any person aggrieved by an order made by the State Board under section 25, section 26 or section 27 may, within thirty days from the date on which the order is communicated to him, prefer an appeal to such authority (hereinafter referred to as the Appellate Authority) as the State Government may think fit to constitute:

PROVIDED that the Appellate Authority may entertain the appeal after the expiry of the said period of thirty days if such authority is satisfied that the appellant was prevented by sufficient cause from filing the appeal in time.

(2) [An Appellate Authority shall consist of a single person or three persons as the State Government may think fit, to be appointed by that government.][13]

(3) The form and manner in which an appeal may be preferred under sub-section (1), the fees payable for such appeal and the procedure to be followed by the Appellate Authority shall be such as may be prescribed.

(4) On receipt of an appeal preferred under sub-section (1), the Appellate Authority shall, after giving the appellant and the State Board an opportunity of being heard, dispose of the appeal as expeditiously as possible.

(5) If the Appellate Authority determines that any condition imposed, or the variation of any condition, as the case may be, was unreasonable, then-

(a) where the appeal is in respect of the unreasonableness of any condition imposed, such authority may direct either that the condition shall be treated as annulled or that there shall be substituted for it such condition as appears to it to be reasonable;

(b) where the appeal is in respect of the unreasonableness of any variation of a condition, such authority may direct either that the condition shall be treated as continuing in force unvaried or that it shall be varied in such manner as appears to it to be reasonable.

29. Revision

(1) The State Government may at any time either of its own motion or on an application made to it in this behalf, call for the records of any case where an order has been made by the State Board under section 25, section 26 or section 27 for the purpose of satisfying itself as to the legality or propriety of any such order and may pass such order in relation thereto as it may think it:

PROVIDED that the State Government shall not pass any order under this sub-section without affording the State Board and the person who may be affected by such order a reasonable opportunity of being heard in the matter.

(2) The State Government shall not revise any order made under section 25, section 26, or section 27 where an appeal against that order lies to the Appellate Authority, but has not been preferred or where an appeal has been preferred such appeal is pending before the Appellate Authority.

30. Power of State Board to Carry Out Certain Works

(1) Where under this Act, any conditions have been imposed on any person while granting consent under section 25 or section 26 and such conditions require such person to execute any work in connection therewith and such work has not been executed within such time as may be specified in this behalf, the State Board may serve on the person concerned a notice requiring him within such time (not being less than thirty days) as may be specified in the notice to execute the work specified therein.

(2) If the person concerned fails to execute the work as required in the notice referred to in sub-section (1), then, after the expiration of the time specified in the said notice, the State Board may itself execute or cause to be executed such work.

(3) All expenses incurred by the State Board for the execution of the aforesaid work, together with interest, at such rate as the State Government may, by order, fix, from the date when a demand for the expenses is made until it is paid, may be recovered by that Board from the person concerned, as arrears of land revenue, or of public demand.

31. Furnishing of Information to State Board and Other Agencies in Certain Cases

(1) If at any place where any industry, operation or process, or any treatment and disposal system or any extension or addition thereto is being carried on, due to accident or other unforeseen act or event, any poisonous, noxious or polluting matter is being discharged, or is likely to be discharged into a stream or well or sewer or on land and, as a result of such discharge, the water in any stream or well is being

polluted, or is likely to be polluted, then the person in charge of such place shall forthwith intimate the occurrence of such accident, act or event to the State Board and such other authorities or agencies as may be prescribed.

(2) Where any local authority operates any sewerage system or sewage works, the provisions of sub-section (1) shall apply to such local authority as they apply in relation to the person in charge of the place where any industry or trade is being carried on.

32. Emergency Measures in Case of Pollution of Stream or Well

(1) Where it appears to the State Board that any poisonous, noxious or polluting matter is present in [any stream or well or on land by reason of the discharge of such matter in such stream or well or on such land] or has entered into that stream or well due to any accident or other unforeseen act or event, and if the Board is of opinion that it is necessary or expedient to take immediate action, it may for reasons to be recorded in writing, carry out such operations, as it may consider necessary for all or any of the following purposes, that is to say-

(a) removing that matter from the [stream or well or on land][21] and disposing it of in such manner as the Board considers appropriate;

(b) remedying or mitigating any pollution caused by its presence in the stream or well;

(c) issuing orders immediately restraining or prohibiting the person concerned from discharging any poisonous, noxious or polluting matter [into the stream or well or on land] or from making insanitary use of the stream or well.

(2) The power conferred by sub-section (1) does not include the power to construct any works other than works of a temporary character which are removed on or before the completion of the operation.

33. Power of Board to Make Application to Courts for Restraining Apprehended Pollution of Water in Streams or Wells

(1) Where it is apprehended by a Board that the water in any stream or well is likely to be polluted by reason of the disposal or likely disposal of any matter in such stream or well or in any sewer or on any land, or otherwise, the Board may make an application to a court, not inferior to that of a Metropolitan Magistrate or a Judicial Magistrate of the first class, for restraining the person who is likely to cause such pollution from so causing.

(2) On receipt of an application under sub-section (1) the court may make such order as it deems fit.

(3) Where under sub-section (2) the court makes an order restraining any person from polluting the water in any stream or well, it may in that order-

(i) direct the person who is likely to cause or has caused the pollution of the water in the stream or well, to desist from taking such action as is likely to cause pollution or, as the case may be, to remove from such stream or well, such matter, and

(ii) authorise the Board, if the direction under clause (i) (being a direction for the removal of any matter from such stream or well) is not complied with by the person to whom such direction is issued, to undertake the removal and disposal of the matter in such manner as may be specified by the court.

(4) All expenses incurred by the Board in removing any matter in pursuance of the authorisation under clause (ii) of sub-section (3) or in the disposal of any such matter may be defrayed out of any money obtained by the Board from such disposal and any balance outstanding shall be recoverable from the person concerned as arrears of land revenue or of public demand.

33A. Power to Give Directions

Notwithstanding anything contained in any other law, but subject to the provisions of this Act, and to any directions that the Central Government may give in this behalf, a Board may, in the exercise of its powers and performance of its functions under this Act, issue any directions in writing to any person, officer or authority, and such person, officer or authority shall be bound to comply with such directions.

Explanation : For the avoidance of doubts, it is hereby declared that the power to issue directions under this section includes the power to direct-

(a) the closure, prohibition or regulation of any industry, operation or process; or

(b) the storage or regulation of supply of electricity, water or any other service.

CHAPTER VI : FUNDS, ACCOUNTS AND AUDIT

34. Contributions by Central Government

The Central Government may, after due appropriation made by Parliament by law in this behalf, make in each financial year such contributions to the Central Board as it may think necessary to enable the Board to perform its functions under this Act.

35. Contributions by State Government

The State Government may, after due appropriation made by the Legislature of the State by law in this behalf, make in each financial year such contributions to the State Board as it may think necessary to enable that Board to perform its functions under this Act.

36. Fund of Central Board

(1) The Central Board shall have its own fund, and all sums which may, from time to time, be paid to it by the Central Government and all other receipts (by way of gifts, grants, donations, benefactions [fees] or otherwise) of that Board shall be carried to the fund of the Board and all payments by the Board shall be made therefrom.

(2) The Central Board may expend such sums as it thinks fit for performing its functions under this Act, [and, where any law for the time being in force relating to the prevention, control or abatement of air pollution provides for the performance of any function under such law by the Central Board, also for performing its functions under such law] and such sums shall be treated as expenditure payable out of the fund of the Board.

37. Fund of State Board

(1) The State Board shall have its own fund, and the sums which may, from time to time, be paid to it by the State Government and all other receipts (by way of gifts, grants, donations, benefactions [fees] or otherwise) of that Board shall be carried to the fund of the Board and all payments by the Board shall be made therefrom.

(2) The State Board may expend such sums as it thinks fit for performing its functions under this Act, [and,

where any law for the time being in force relating to the prevention, control or abatement of air pollution provides for the performance of any function under such law by the State Board, also for performing its functions under such law] and such sums shall be treated as expenditure payable out of the fund of the Board.

37A. [Borrowing Powers of Board

A Board may, with the consent of, or in accordance with, the terms of any general or special authority given to it by the Central Government or, as the case may be, the State Government, borrow money from any source by way of loans or issue of bonds, debentures or such other instruments, as it may deem fit, for the performance of all or any of its functions under this Act.

38. Budget

The Central Board or, as the case may be, the State Board shall, during each financial year, prepare, in such form and at such time as may be prescribed, a budget in respect of the financial year next ensuing showing the estimated receipt and expenditure, and copies thereof shall be forwarded to the Central Government or, as the case may be, the State Government.

39. Annual Report

(1) The Central Board shall, during each financial year, prepare, in such form as may be prescribed, an annual report giving full account of its activities under this Act during the previous financial year and copies thereof shall be forwarded to the Central Government within four months from the last date of the previous financial year and that government shall cause every

such report to be laid before both Houses of Parliament within nine months from the last date of the previous financial year.

(2) Every State Board shall, during each financial year, prepare, in such form as may be prescribed, an annual report giving full account of its activities under this Act during the previous financial year and copies thereof shall be forwarded to the State Government within four months from the last date of the previous financial year and that government shall cause every such report to be laid before the State Legislature within a period of nine months from the last date of the previous financial year.

40. Accounts and Audit

(1) Every Board shall maintain proper accounts and other relevant records and prepare an annual statement of accounts in such form as may be prescribed by the Central Government or, as the case may be, the State Government.

(2) The accounts of the Board shall be audited by an auditor duly qualified to act as an auditor of companies under section 226 of the Companies Act, 1956 (1 of 1956).

(3) The said auditor shall be appointed by the Central Government or, as the case may be, the State Government on the advice of the Comptroller and Auditor-General of India.

(4) Every auditor appointed to audit the accounts of the Board under this Act shall have the right to demand the production of books, accounts, connected vouchers and other documents and papers and to inspect any of the offices of the Board.

(5) Every such auditor shall send a copy of his report together with an audited copy of the accounts to the Central Government or, as case may be, the State Government.

(6) The Central Government shall, as soon as may be after the receipt of the audit report under sub-section (5), cause the same to be laid before both Houses of Parliament.

(7) The State Government shall, as soon as may be after the receipt of the audit report under sub-section (5), cause the same to be laid before the State Legislature.

CHAPTER VII : PENALTIES AND PROCEDURE

41. [Failure to Comply with Directions Under Sub-Section (2) or Sub-Section (3) of Section 20,

or orders issued under clause (c) of sub-section (1) of section 32 or directions issued under sub-section (2) of section 33 or section 33A

(1) Whoever fails to comply with the direction given under sub-section (2) or sub-section (3) of section 20 within such time as may be specified in the direction shall, on conviction, be punishable with imprisonment for a term which may extend to three months or with fine which may extend to ten thousand rupees or with both and in case the failure continues, with an additional fine which may extend to five thousand rupees for every day during which such failure continues after the conviction for the first such failure.

(2) Whoever fails to comply with any order issued under clause (e) of sub-section (1) of section 32 or any direction issued by a court under sub-section (2) of section 33 or any direction issued under section 33A

shall, in respect of each such failure and on conviction, be punishable with imprisonment for a term which shall not be less than one year and six months but which may extend to six years and with fine, and in case the failure continues, with an additional fine which may extend to five thousand rupees for everyday during which such failure continues after the conviction for the first such failure.

(3) If the failure referred to in sub-section (2) continues beyond a period of one year after the date of conviction, the offender shall, on conviction, be punishable with imprisonment for a term which shall not be less than two years but which may extend to seven years and with fine.

42. Penalty for Certain Acts

(1) Whoever-

(a) destroys, pulls down, removes, injures or defaces any pillar, post or stake fixed in the ground or any notice or other matter put up, inscribed or placed, by or under the authority of the Board, or

(b) obstructs any person acting under the orders or directions of the Board from exercising his powers and performing his functions under this Act, or

(c) damages any works or property belonging to the Board, or

(d) fails to furnish to any officer or other employee of the Board any information required by him for the purpose of this Act, or

(e) fails to intimate the occurrence of any accident or other unforeseen act or event under section 31 to the Board and other authorities or agencies as required by that section, or

(f) in giving any information which he is required to give under this Act, knowingly or wilfully makes a statement which is false in any material particular, or

(g) for the purpose of obtaining any consent under section 25 or section 26, knowingly or wilfully makes a statement which is false in any material particular,

shall be punishable with imprisonment for a term which may extend to three months or with fine which may extend to [ten thousand rupees] or with both.

(2) Where for the grant of a consent in pursuance of the provisions of section 25 or section 26 the use of meter or gauge or other measure or monitoring device is required and such device is used for the purposes of those provisions, any person who knowingly or wilfully alters or interferes with that device so as to prevent it from monitoring or measuring correctly shall be punishable with imprisonment for a term which may extend to three months or with fine which may extend to [ten thousand rupees] or with both.

43. Penalty for Contravention of Provisions of Section 24

Whoever contravenes the provisions of section 24 shall be punishable with imprisonment for a term which shall not be less than [one year and six months] but which may extend to six years and with fine.

44. Penalty for Contravention of Section 25 or Section 26

Whoever contravenes the provisions of section 25 or section 26 shall be punishable with imprisonment for a term which shall not be less than [one year and six months] but which may extend to six years and with fine.

45. Enhanced Penalty After Previous Conviction

If any person who has been convicted of any offence under section 24 or section 25 or section 26 is again found guilty of an offence involving a contravention of the same provision, he shall, on the second and on every subsequent conviction, be punishable with imprisonment for a term which shall not be less than [two years] but which may extend to seven years and with fine:

PROVIDED that for the purpose of this section no cognizance shall be taken of any conviction made more than two years before the commission of the offence which is being punished.

45A. Penalty for contravention of certain provisions of the Act

Whoever contravenes any of the provisions of this Act or fails to comply with any order or direction given under this Act, for which no penalty has been elsewhere provided in this Act, shall be punishable with imprisonment which may extend to three months or with fine which may extend to ten thousand rupees or with both and in the case of a continuing contravention or failure, with an additional fine which may extend to five thousand rupees for every day during which such contravention or failure continues after conviction for the first such contravention or failure.]

46. Publication of names of Offenders

If any person convicted of an offence under this Act commits a like offence afterwards it shall be lawful for the court before which the second or subsequent conviction takes place to cause the offender's name and place of residence, the offence and the penalty imposed to be published at the offender's expense in such newspapers or in such other manner as the court may direct and the expenses of such publication shall be deemed to be part of the cost attending the conviction and shall be recoverable in the same manner as a fine.

47. Offences by Companies

(1) Where an offence under this Act has been committed by a company, every person who at the time the offence was committed was in charge of, and was responsible to the company for the conduct of, the business of the company, as well as the company, shall be deemed to be guilty of the offence and shall be liable to be proceeded against and punished accordingly:

PROVIDED that nothing contained in this sub-section shall render any such person liable to any punishment provided in this Act if he proves that the offence was committed without his knowledge or that he exercised all due diligence to prevent the commission of such offence.

(2) Notwithstanding anything contained in sub-section (1), where an offence under this Act has been committed by a company and it is proved that the offence has been committed with the consent or connivance of, or is attributable to any neglect on the part of, any director, manager, secretary or other officer of the company, such director, manager, secretary or other officer shall also be deemed to be guilty of that offence and shall be liable to be proceeded against and punished accordingly.

Explanation : For the purposes of this section-

(a) "company" means any body corporate, and includes a firm or other association of individuals; and

(b) "director" in relation to a firm means a partner in the firm.

48. Offences by Government Departments

Where an offence under this Act has been committed by any Department of Government, the Head of the Department shall be deemed to be guilty of the offence and shall be liable to be proceeded against and punished accordingly.

PROVIDED that nothing contained in this section shall render such Head of the Department liable to any punishment if he proves that the offence was committed without his knowledge or that he exercised all due diligence to prevent the commission of such offence.

49. Cognizance of Offences

(1) No court shall take cognizance of any offence under this Act except on a complaint made by-

(a) a Board or any officer authorised in this behalf by it; or

(b) any person who has given notice of not less than sixty days, in the manner prescribed, of the alleged offence and of his intention to make a complaint, to the Board or officer authorised as aforesaid, and no court inferior to that of a Metropolitan Magistrate or a Judicial Magistrate of the first class shall try any offence punishable under this Act.

(2) [Where a complaint has been made under clause (b) of sub-section (1), the Board shall, on demand by such person, make available the relevant reports in its possession to that person:

PROVIDED that the Board may refuse to make any such report available to such person if the same is in its opinion, against the public interest.

(3) Notwithstanding anything contained in section 29 of the Code of Criminal Procedure, 1973 (2 of 1974), it shall be lawful for any [Judicial Magistrate of the first class or for any Metropolitan Magistrate] to pass a sentence of imprisonment for a term exceeding two years or of fine exceeding two thousand rupees on any person convicted of an offence punishable under this Act.

50. Members, Officers and Servants of Board to be Public Servants

All members, officers and servants of a Board when acting or purporting to act in pursuance of any of the provisions of this Act and the rules made thereunder shall be deemed to be public servants within the meaning of section 21 of the Indian Penal Code (45 of 1860).

CHAPTER VIII : MISCELLANEOUS

51. Central Water Laboratory

(1) The Central Government may, by notification in the Official Gazette-

(a) establish a Central Water Laboratory; or

(b) specify any laboratory or institute as a Central Water Laboratory, to carry out the functions entrusted to the Central Water Laboratory under this Act.

(2) The Central Government may, after consultation with the Central Board, make rules prescribing-

(a) the functions of the Central Water Laboratory;

(b) the procedure for the submission to the said laboratory of samples of water or of sewage or trade effluent for analysis or tests, the form of the laboratory's report thereunder and the fees payable in respect of such report;

(c) such other matters as may be necessary or expedient to enable that laboratory to carry out its functions.

52. State Water Laboratory

(1) The State Government may, by notification in the Official Gazette-

(a) establish a State Water Laboratory; or

(b) specify any laboratory or institute as a State Water Laboratory, to carry out the functions entrusted to the State Water Laboratory under this Act.

(2) The State Government may, after consultation with the State Board, make rules prescribing-

(a) the functions of the State Water Laboratory;

(b) the procedure for the submission to the said laboratory of samples of water or of sewage or trade effluent for analysis or tests, the form of the laboratory's report thereon and the fees payable in respect of such report;

(c) such other matters as may be necessary or expedient to enable that laboratory to carry out its functions.

53. Analysts

(1) The Central Government may, by notification in the Official Gazette, appoint such persons as it thinks fit and having the prescribed qualifications to be

government analysts for the purpose of analysis of samples of water or of sewage or trade effluent sent for analysis to any laboratory established or specified under sub-section (1) of section 51.

(2) The State Government may, by notification in the Official Gazette, appoint such persons as it thinks fit and having the prescribed qualifications to be government analysts for the purpose of analysis of samples of water or of sewage or trade effluent sent for analysis to any laboratory established or specified under sub-section (1) of section 52.

(3) Without prejudice to the provisions of sub-section (3) of section 12, the Central Board or, as the case may be, the State Board may, by notification in the Official Gazette, and with the approval of the Central Government or the State Government, as the case may be, appoint such persons as it thinks fit and having the prescribed qualifications to be Board analysts for the purpose of analysis of samples of water or of sewage or trade effluent sent for analysis to any laboratory established or recognised under section 16, or, as the case may be, under section 17.

54. Reports of Analysts

Any document purporting to be a report signed by a government analyst or, as the case may be, a Board analyst may be used as evidence of the facts stated therein in any proceeding under this Act.

55. Local Authorities to Assist

All local authorities shall render such help and assistance and furnish such information to the Board as it may require for the discharge of its functions, and shall make available to the Board for inspection and examination such records, maps, plans and other documents as may be necessary for the discharge of its functions.

56. Compulsory Acquisition of Land for the State Board

Any land required by a State Board for the efficient performance of its functions under this Act shall be deemed to be needed for a public purpose and such land shall be acquired for the State Board under the provisions of the Land Acquisition Act, 1894 (1 of 1894), or under any other corresponding law for the time being in force:

57. Returns and Reports

The Central Board shall furnish to the Central Government, and a State Board shall furnish to the State Government and to the Central Board such reports, returns, statistics, accounts and other information with respect to its fund or activities as that government, or, as the case may be, the Central Board may, from time to time, require.

58. Bar of Jurisdiction

No civil court shall have jurisdiction to entertain any suit or proceeding in respect of any matter which an Appellate Authority constituted under this Act is empowered by or under this Act to determine, and no injunction shall be granted by any court or other authority in respect of any action taken or to be taken in pursuance of any power conferred by or under this Act.

59. Protection of Action Taken in Good Faith

No suit or other legal proceedings shall lie against the government or any officer of government or any member or officer of a Board in respect of anything which is in good faith done or intended to be done in pursuance of this Act or the rules made thereunder.

60. Overriding Effect

The provisions of this Act shall have effect notwithstanding anything inconsistent therewith contained in any enactment other than this Act.

61. Power of Central Government to Supersede the Central Board and Joint Boards

(1) If at any time the Central Government is of opinion-

(a) that the Central Board or any Joint Board has persistently made default in the performance of the functions imposed on it by or under this Act; or

(b) that circumstances exist which render it necessary in the public interest so to do, the Central Government may, by notification in the Official Gazette, supersede the Central Board or such Joint Board, as the case may be, for such period, not exceeding one year, as may be specified in the notification:

PROVIDED that before issuing a notification under this sub-section for the reasons mentioned in clause (a), the Central Government shall give a reasonable opportunity to the Central Board or such Joint Board, as the case may be, to show cause why it should not be superseded and shall consider the explanations and objections if any, of the Central Board or such Joint Board, as the case may be.

(2) Upon the publication of a notification under sub-section (1) superseding the Central Board or any Joint Board-

(a) all the members shall, as from the date of supersession vacate their offices as such;

(b) all the powers, functions and duties which may, by or under this Act, be exercised, performed or discharged by the Central Board or such Joint board shall, until the Central Board or the Joint Board, as the case may be, is reconstituted under sub-section (3) be exercised, performed or discharged by such person or persons as the Central Government may direct;

(c) all property owned or controlled by the Central Board or such Joint Board shall, until the Central Board or the Joint Board, as the case may be, is reconstituted under sub-section (3) vest in the Central Government.

(3) On the expiration of the period of supersession specified in the notification issued under sub-section (1), the Central Government may-

(a) extend the period of supersession for such further term, not exceeding six months, as it may consider necessary; or

(b) reconstitute the Central Board or the Joint Board, as the case may be, by fresh nomination or appointment, as the case may be, and in such case any person who vacated his office under clause (a) of sub-section (2) shall not be deemed disqualified for nomination or appointment:

PROVIDED that the Central Government may at any time before the expiration of the period of supersession, whether originally specified under sub-section (1) or as extended under this sub-section, take action under clause (b) of this sub-section.

62. Power of State Government to Supersede State Board

(1) If at any time the State Government is of opinion-

(a) that the State Board has persistently made default in the performance of the functions imposed on it by or under this Act; or

(b) that circumstances exist which render it necessary in the public interest so to do, the State Government may, by notification in the Official Gazette, supersede the State Board for such period, not exceeding one year, as may be specified in the notification:

PROVIDED that before issuing a notification under this sub-section for the reasons mentioned in clause (a), the State Government shall give a reasonable opportunity to the State Board to show cause why it should not be superseded and shall consider the explanations and objections, if any, of the State Board.

(2) Upon the publication, of a notification under sub-section (1) superseding the State Board, the provisions of sub-sections (2) and (3) of section 61 shall apply in relation to the supersession of the State Board as they apply in relation to the supersession of the Central Board or a Joint Board by the Central Government.

63. Power of Central Government to make rules

(1) The Central Government may, simultaneously with the constitution of the Central Board, make rules in respect of the matters specified in sub-section (2):

PROVIDED that when the Central Board has been constituted, no such rule shall be made, varied, amended or repealed without consulting the Board.

(2) In particular, and without prejudice to the generality of the foregoing power, such rules may provide for all or any of the following matters, namely,-

(a) the terms and conditions of service of the members (other than the Chairman and member-secretary) of the Central Board under sub-section (8) of section 5;

(b) the intervals and the time and place at which meetings of the Central Board or of any committee thereof constituted under this Act, shall be held and the procedure to be followed at such meetings, including the quorum necessary for the transaction of business under section 8, and under sub-section (2) of section 9;

(c) the fees and allowances to be paid to such members of a committee of the Central Board as are not members of the Board under sub-section (3) of section 9;

(d) [the manner in which and the purposes for which persons may be associated with the Central Board under sub-section (1) of section 10 and the fees and allowances payable to such persons;]

(e) the terms and conditions of service of the Chairman and the member-secretary of the Central Board under sub-section (9) of section 5 and under sub-section (1) of section 12;

(f) conditions subject to which a person may be appointed as a consulting engineer to the Central Board under sub-section (4) of section 12;

(g) the powers and duties to be exercised and performed by the Chairman and the member-secretary of the Central Board;

[(h) ***; (i) ***]

(j) the form of the report of the Central Board analyst under sub-section (1) of section 22;

(k) the form of the report of the government analyst under sub-section (3) of section 22;

(l) [the form in which and the time within which the budget of the Central Board may be prepared and forwarded to the Central Government under section 38;

(i) the form in which the annual report of the Central Board may be prepared under section 39;]

(m) the form in which the accounts of the Central Board may be maintained under section 40;

(i) [the manner in which notice of intention to make a complaint shall be given to the Central Board or officer authorised by it under section 49;]

(n) any other matter relating to the Central Board, including the powers and functions of that Board in relation to Union Territories;

(o) any other matter which has to be, or may be, prescribed.

(3) Every rule made by the Central Government under this Act shall be laid, as soon as may be after it is made, before each House of Parliament while it is in session for a total period of thirty days which may be comprised in one session or in two or more successive sessions, and if, 5[before the expiry of the session immediately following the session or the successive sessions aforesaid], both Houses agree in making any modification in the rule or both Houses agree that the rule should not be made, the rule shall thereafter have effect only in such modified form or be of no effect, as the case may be; so, however, that any such modification or annulment shall be without prejudice to the validity of anything previously done under that rule.

64. Power of State Government to Make Rules

(1) The State Government may, simultaneously with the constitution of the State Board, make rules to carry out the purposes of this Act, in respect of matters not falling within the purview of section 63:

PROVIDED that when the State Board has been constituted, no such rule shall be made, varied, amended or repealed without consulting that Board.

(2) In particular, and without prejudice to the generality of the foregoing power, such rules may provide for all or any of the following matters, namely,-

(a) the terms and conditions of service of members (other than the Chairman and the member-secretary) of the State Board under sub-section (8) of section 5;

(b) the time and place of meetings of the State Board or of any committee of that Board constituted under this Act and the procedure to be followed at such meeting, including the quorum necessary for the transaction of business under section 8 and under sub-section (2) of section 9;

(c) the fees and allowances to be paid to such members of a committee of the State Board as are not members of the Board under sub-section (3) of section 9;

(d) the manner in which and the purposes for which persons may be associated with the State Board under sub-section (1) of section 10 [and the fees and allowances payable to such persons;]

(e) the terms and conditions of service of the Chairman and the member-secretary of the State Board under sub-section (9) of section 5 and under sub-section (1) of section 12;

(f) the conditions subject to which a person may be appointed as a consulting engineer to the State Board under sub-section (4) of section 12;

(g) the powers and duties to be exercised and discharged by the Chairman and the member-secretary of the State Board;

(h) the form of the notice referred to in section 21;

(i) the form of the report of the State Board analyst under sub-section (1) of section 22;

(j) the form of the report of the government analyst under sub-section (3) of section 22;

(k) the form of application for the consent of the State Board under sub-section (2) of section 25, and the particulars it may contain;

(l) the manner in which inquiry under sub-section (3) of section 25 may be made in respect of an application for obtaining consent of the State Board and the matters to be taken into account in granting or refusing such consent;

(m) the form and manner in which appeals may be filed, the fees payable in respect of such appeals and the procedure to be followed by the Appellate Authority in disposing of the appeals under sub-section(3) of section 28;

(n) [the form in which and the time within which the budget of the State Board may be prepared and forwarded to the State Government under section 38;

(i) the form in which the annual report of the State Board may be prepared under section 39;

(o) the form in which the accounts of the State Board may be maintained under sub-section (1) of section 40;

(i) the manner in which notice of intention to make a complaint shall be given to the State Board or officer authorised by it under section 49;

(p) any other matter which has to be, or may be, prescribed.

ANNEXURE III

The Environment (Protection) Act, 1986

CHAPTER I : PRELIMINARY

1. Short Title, Extend and Commencement

(1) This Act may be called the Environment (Protection) Act, 1986.

(2) It extends to the whole of India.

(3) It shall come into force on such date as the Central Government may, by notification in the Official Gazette, appoint and different dates may be appointed for different provisions of this Act and for different areas.

2. Definitions

In this Act, unless the context otherwise requires,—

(a) "environment" includes water, air and land and the inter-relationship which exists among and between water, air and land, and human beings, other living creatures, plants, micro-organism and property;

(b) "environmental pollutant" means any solid, liquid or gaseous substance present in such concentration as may be, or tend to be, injurious to environment;

(c) "environmental pollution" means the presence in the environment of any environmental pollutant;

(d) "handling", in relation to any substance, means the manufacture, processing, treatment, package, storage, transportation, use, collection, destruction, conversion, offering for sale, transfer or the like of such substance;

(e) "hazardous substance" means any substance or preparation which, by reason of its chemical or physico-chemical properties or handling, is liable to cause harm to human beings, other living creatures, plant, micro-organism, property or the environment;

(f) "occupier", in relation to any factory or premises, means a person who has, control over the affairs of the factory or the premises and includes in relation to any substance, the person in possession of the substance;

(g) "prescribed" means prescribed by rules made under this Act.

CHAPTER II. GENERAL POWERS OF THE CENTRAL GOVERNMENT

3. Power of Central Government to Take Measures to Protect and Improve Environment

(1) Subject to the provisions of this Act, the Central Government, shall have the power to take all such measures as it deems necessary or expedient for the purpose of protecting and improving the quality of the environment and preventing controlling and abating environmental pollution.

(2) In particular, and without prejudice to the generality of the provisions of sub-section (1), such measures may include measures with respect to all or any of the following matters, namely:—

(i) co-ordination of actions by the State Governments, officers and other authorities—

(a) under this Act, or the rules made thereunder, or

(b) under any other law for the time being in force which is relatable to the objects of this Act;

(ii) planning and execution of a nation-wide programme for the prevention, control and abatement of environmental pollution;

(iii) laying down standards for the quality of environment in its various aspects;

(iv) laying down standards for emission or discharge of environmental pollutants from various sources whatsoever:

Provided that different standards for emission or discharge may be laid down under this clause from different sources having regard to the quality or composition of the emission or discharge of environmental pollutants from such sources;

(v) restriction of areas in which any industries, operations or processes or class of industries, operations or processes shall not be carried out or shall be carried out subject to certain safeguards;

(vi) laying down procedures and safeguards for the prevention of accidents which may cause environmental pollution and remedial measures for such accidents;

(vii) laying down procedures and safeguards for the handling of hazardous substances;

(viii) examination of such manufacturing processes, materials and substances as are likely to cause environmental pollution;

(ix) carrying out and sponsoring investigations and research relating to problems of environmental pollution;

(x) inspection of any premises, plant, equipment, machinery, manufacturing or other processes, materials or substances and giving, by order, of such directions to such authorities, officers or persons as it may consider necessary to take steps for the prevention, control and abatement of environmental pollution;

(xi) establishment or recognition of environmental laboratories and institutes to carry out the functions entrusted to such environmental laboratories and institutes under this Act;

(xii) collection and dissemination of information in respect of matters relating to environmental pollution;

(xiii) preparation of manuals, codes or guides relating to the prevention, control and abatement of environmental pollution;

(xiv) such other matters as the Central Government deems necessary or expedient for the purpose of securing the effective implementation of the provisions of this Act.

(3) The Central Government may, if it considers it necessary or expedient so to do for the purpose of this Act, by order, published in the Official Gazette, constitute an authority or authorities by such name or names as may be specified in the order for the purpose of exercising and performing such of the powers and functions (including the power to issue

directions under section 5) of the Central Government under this Act and for taking measures with respect to such of the matters referred to in sub-section (2) as may be mentioned in the order and subject to the supervision and control of the Central Government and the provisions of such order, such authority or authorities may exercise and powers or perform the functions or take the measures so mentioned in the order as if such authority or authorities had been empowered by this Act to exercise those powers or perform those functions or take such measures.

4. Appointment of Officers and their Powers and Functions

(1) Without prejudice to the provisions of sub-section (3) of section 3, the Central Government may appoint officers with such designation as it thinks fit for the purposes of this Act and may entrust to them such of the powers and functions under this Act as it may deem fit.

(2) The officers appointed under sub-section (1) shall be subject to the general control and direction of the Central Government or, if so directed by that Government, also of the authority or authorities, if any, constituted under sub- section (3) of section 3 or of any other authority or officer.

5. Power to Give Directions

Notwithstanding anything contained in any other law but subject to the provisions of this Act, the Central Government may, in the exercise of its powers and performance of its functions under this Act, issue directions in writing to any person, officer or any authority and such person, officer or authority shall be bound to comply with such directions.

Explanation – For the avoidance of doubts, it is hereby declared that the power to issue directions under this section includes the power to direct–

(a) the closure, prohibition or regulation of any industry, operation or process; or

(b) stoppage or regulation of the supply of electricity or water or any other service.

6. Rules to Regulate Environmental Pollution

(1) The Central Government may, by notification in the Official Gazette, make rules in respect of all or any of the matters referred to in section 3.

(2) In particular, and without prejudice to the generality of the foregoing power, such rules may provide for all or any of the following matters, namely:–

(a) the standards of quality of air, water or soil for various areas and purposes;

(b) the maximum allowable limits of concentration of various environmental pollutants (including noise) for different areas;

(c) the procedures and safeguards for the handling of hazardous substances;

(d) the prohibition and restrictions on the handling of hazardous substances in different areas;

(e) the prohibition and restriction on the location of industries and the carrying on process and operations in different areas;

(f) the procedures and safeguards for the prevention of accidents which may cause environmental pollution and for providing for remedial measures for such accidents.

CHAPTER III : PREVENTION, CONTROL, AND ABATEMENT OF ENVIRONMENTAL POLLUTION

7. Persons Carrying on Industry Operation, Etc., Not to Allow Emission or Discharge of Environmental Pollutants in Excess of The Standards

No person carrying on any industry, operation or process shall discharge or emit or permit to be discharged or emitted any environmental pollutants in excess of such standards as may be prescribed.

8. Persons Handling Hazardous Substances to Comply with Procedural Safeguards

No person shall handle or cause to be handled any hazardous substance except in accordance with such procedure and after complying with such safeguards as may be prescribed.

9. Furnishing of Information to Authorities and Agencies in Certain Cases

(1) Where the discharge of any environmental pollutant in excess of the prescribed standards occurs or is apprehended to occur due to any accident or other unforeseen act or event, the person responsible for such discharge and the person in charge of the place at which such discharge occurs or is apprehended to occur shall be bound to prevent or mitigate the environmental pollution caused as a result of such discharge and shall also forthwith—

(a) intimate the fact of such occurrence or apprehension of such occurrence; and

(b) be bound, if called upon, to render all assistance, to such authorities or agencies as may be prescribed.[11]

(2) On receipt of information with respect to the fact or apprehension on any occurrence of the nature referred to in sub-section (1), whether through intimation under that sub-section or otherwise, the authorities or agencies referred to in sub-section (1) shall, as early as practicable, cause such remedial measures to be taken as necessary to prevent or mitigate the environmental pollution.

(3) The expenses, if any, incurred by any authority or agency with respect to the remedial measures referred to in sub-section (2), together with interest (at such reasonable rate as the Government may, by order, fix) from the date when a demand for the expenses is made until it is paid, may be recovered by such authority or agency from the person concerned as arrears of land revenue or of public demand.

10. Powers of Entry and Inspection

(1) Subject to the provisions of this section, any person empowered by the Central Government in this ehalf shall have a right to enter, at all reasonable times with such assistance as he considers necessary, any place—

(a) for the purpose of performing any of the functions of the Central Government entrusted to him;

(b) for the purpose of determining whether and if so in what manner, any such functions are to be performed or whether any provisions of this Act or the rules made thereunder orany notice, order, direction or authorisation served, made, given or granted under this Act is being or has been complied with;

(c) for the purpose of examining and testing any equipment, industrial plant, record, register, document or any other material object or for conducting a search of any building in which he has reason to believe that an offence under this Act or the rules made thereunder has been or is being or is about to be committed and for seizing any such equipment, industrial plant, record, register, document or other material object if he has reason to believe that it may furnish evidence of the commission of an offence punishable under this Act or the rules made thereunder or that such seizure is necessary to prevent or mitigate environmental pollution.

(2) Every person carrying on any industry, operation or process of handling any hazardous substance shall be bound to render all assistance to the person empowered by the Central Government under sub-section (1) for carrying out the functions under that sub-section and if he fails to do so without any reasonable cause or excuse, he shall be guilty of an offence under this Act.

(3) If any person wilfully delays or obstructs any persons empowered by the Central Government under sub-section (1) in the performance of his functions, he shall be guilty of an offence under this Act.

(4) The provisions of the Code of Criminal Procedure, 1973, or, in relation to the State of Jammu and Kashmir, or an area in which that Code is not in

force, the provisions of any corresponding law in force in that State or area shall, so far as may be, apply to any search or seizures under this section as they apply to any search or seizure made under the authority of a warrant issued under section 94 of the said Code or as the case may be, under the corresponding provision of the said law.

11. Power to Take Sample and Procedure to Be Followed in Connection Therewith

(1) The Central Government or any officer empowered by it in this behalf,[13] shall have power to take, for the purpose of analysis, samples of air, water, soil or other substance from any factory, premises or other place in such manner as may be prescribed.

(2) The result of any analysis of a sample taken under sub-section (1) shall not be admissible in evidence in any legal proceeding unless the provisions of sub-sections (3) and (4) are complied with.

(3) Subject to the provisions of sub-section (4), the person taking the sample under sub-section (1) shall—

(a) serve on the occupier or his agent or person in charge of the place, a notice, then and there, in such form as may be prescribed, of his intention to have it so analysed;

(b) in the presence of the occupier of his agent or person, collect a sample for analysis;

(c) cause the sample to be placed in a container or containers which shall be marked and sealed and shall also be signed both by the person taking the sample and the occupier or his agent or person;

(d) send without delay, the container or the containers to the laboratory established or recognised by the Central Government under section 12.

(4) When a sample is taken for analysis under sub-section (1) and the person taking the sample serves on the occupier or his agent or person, a notice under clause (a) of sub-section (3), then,—

(a) in a case where the occupier, his agent or person wilfully absents himself, the person taking the sample shall collect the sample for analysis to be placed in a container or containers which shall be marked and sealed and shall also be signed by the person taking the sample, and

(b) in a case where the occupier or his agent or person present at the time of taking the sample refuses to sign the marked and sealed container or containers of the sample as required under clause (c) of sub-section (3), the marked and sealed container or containers shall be signed by the person taking the samples, and the container or containers shall be sent without delay by the person taking the sample for analysis to the laboratory established or recognised under section 12 and such person shall inform the Government Analyst appointed or recognised under section 12 in writing, about the wilfull absence of the occupier or his agent or person, or, as the case may be, his refusal to sign the container or containers.

12. Environmental Laboratories

(1) The Central Government may, by notification in the Official Gazette,—

(a) establish one or more environmental laboratories;

(b) recognise one or more laboratories or institutes as environmental laboratories to carry out the functions entrusted to an environmental laboratory under this Act.

(2) The Central Government may, by notification in the Official Gazette, make rules specifying—

(a) the functions of the environmental laboratory;

(b) the procedure for the submission to the said laboratory of samples of air, water, soil or other substance for analysis or tests, the form of the laboratory report thereon and the fees payable for such report;

(c) such other matters as may be necessary or expedient to enable that laboratory to carry out its functions.

13. Government Analysts

The Central Government may by notification in the Official Gazette, appoint or recognise such persons as it thinks fit and having the prescribed qualifications to be Government Analysts for the purpose of analysis of samples of air, water, soil or other substance sent for analysis to any environmental laboratory established or recognised under sub-section (1) of section 12.

14. Reports of Government Analysts

Any document purporting to be a report signed by a Government analyst may be used as evidence of the facts stated therein in any proceeding under this Act.

15. Penalty for Contravention of the Provisions of the Act and the Rules, Orders and Directions

(1) Whoever fails to comply with or contravenes any of the provisions of this Act, or the rules made or orders or directions issued thereunder, shall, in respect of each such failure or contravention, be punishable with imprisonment for a term which may extend to five years with fine which may extend to one lakh

rupees, or with both, and in case the failure or contravention continues, with additional fine which may extend to five thousand rupees for every day during which such failure or contravention continues after the conviction for the first such failure or contravention.

(2) If the failure or contravention referred to in sub-section (1) continues beyond a period of one year after the date of conviction, the offender shall be punishable with imprisonment for a term which may extend to seven years.

16. Offences by Companies

(1) Where any offence under this Act has been committed by a company, every person who, at the time the offence was committed, was directly in charge of, and was responsible to, the company for the conduct of the business of the company, as well as the company, shall be deemed to be guilty of the offence and shall be liable to be proceeded against and punished accordingly:

Provided that nothing contained in this sub-section shall render any such person liable to any punishment provided in this Act, if he proves that the offence was committed without his knowledge or that he exercised all due diligence to prevent the commission of such offence.

(2) Notwithstanding anything contained in sub-section (1), where an offence under this Act has been committed by a company and it is proved that the offence has been committed with the consent or connivance of, or is attributable to any neglect on the part of, any director, manager, secretary or other officer of the company, such director, manager, secretary or other officer shall also deemed to be

guilty of that offence and shall be liable to be proceeded against and punished accordingly.

Explanation – For the purpose of this section,–

(a) "company" means any body corporate and includes a firm or other association of individuals;

(b) "director", in relation to a firm, means a partner in the firm.

17. Offences by Government Departments

(1) Where an offence under this Act has been committed by any Department of Government, the Head of the Department shall be deemed to be guilty of the offence and shall be liable to be proceeded against and punished accordingly.

Provided that nothing contained in this section shall render such Head of the Department liable to any punishment if he proves that the offence was committed without his knowledge or that he exercise all due diligence to prevent the commission of such offence.

(2) Notwithstanding anything contained in sub-section (1), where an offence under this Act has been committed by a Department of Government and it is proved that the offence has been committed with the consent or connivance of, or is attributable to any neglect on the part of, any officer, other than the Head of the Department, such officer shall also be deemed to be guilty of that offence and shall be liable to be proceeded against and punished accordingly.

CHAPTER IV : MISCELLANEOUS

18. Protection of Action Taken in Good Faith

No suit, prosecution or other legal proceeding shall lie against the Government or any officer or other employee of the Government or any authority constituted under this Act or any member, officer or other employee of such authority in respect of anything which is done or intended to be done in good faith in pursuance of this Act or the rules made or orders or directions issued thereunder.

19. Cognizance of Offences

No court shall take cognizance of any offence under this Act except on a complaint made by—

(a) the Central Government or any authority or officer authorised in this behalf by that Government, or

(b) any person who has given notice of not less than sixty days, in the manner prescribed, of the alleged offence and of his intention to make a complaint, to the Central Government or the authority or officer authorised as aforesaid.

20. Information, Reports or Returns

The Central Government may, in relation to its function under this Act, from time to time, require any person, officer, State Government or other authority to furnish to it or any prescribed authority or officer any reports, returns, statistics, accounts and other information and such person, officer, State Government or other authority shall be bound to do so.

21. Members, Officers and Employees of the Authority Constituted Under Section 3 to be Public Servants

All the members of the authority, constituted, if any, under section 3 and all officers and other employees of such authority when acting or purporting to act in pursuance of any provisions of this Act or the rules made or orders or directions issued thereunder shall be deemed to be public servants within the meaning of section 21 of the Indian Penal Code (45 of 1860).

22. Bar of Jurisdiction

No civil court shall have jurisdiction to entertain any suit or proceeding in respect of anything done, action taken or order or direction issued by the Central Government or any other authority or officer in pursuance of any power conferred by or in relation to its or his functions under this Act.

23. Powers to Delegate

Without prejudice to the provisions of sub-section (3) of section 3, the Central Government may, by notification in the Official Gazette, delegate, subject to such conditions and limitations as may be specified in the notifications, such of its powers and functions under this Act [except the powers to constitute an authority under sub-section (3) of section 3 and to make rules under section 25] as it may deem necessary or expedient, to any officer, State Government or other authority.

24. Effect of Other Laws

(1) Subject to the provisions of sub-section (2), the provisions of this Act and the rules or orders made therein shall have effect notwithstanding anything inconsistent therewith contained in any enactment other than this Act.

(2) Where any act or omission constitutes an offence punishable under this Act and also under any other Act then the offender found guilty of such offence shall be liable to be punished under the other Act and not under this Act.

25. Power to Make Rules

(1) The Central Government may, by notification in the Official Gazette, make rules for carrying out the purposes of this Act.

(2) In particular, and without prejudice to the generality of the foregoing power, such rules may provide for all or any of the following matters, namely—

(a) the standards in excess of which environmental pollutants shall not be discharged or emitted under section 7;

(b) the procedure in accordance with and the safeguards in compliance with which hazardous substances shall be handled or caused to be handled under section 8;

(c) the authorities or agencies to which intimation of the fact of occurrence or apprehension of occurrence of the discharge of any environmental pollutant in excess of the prescribed standards shall be given and to whom all assistance shall be bound to be rendered under sub-section (1) of section 9;

(d) the manner in which samples of air, water, soil or other substance for the purpose of analysis shall be taken under sub-section (1) of section 11;

(e) the form in which notice of intention to have a sample analysed shall be served under clause (a) of sub section (3) of section 11;

(f) the functions of the environmental laboratories, the procedure for the submission to such laboratories of samples of air, water, soil and other substances for analysis or test; the form of laboratory report; the fees payable for such report and other matters to enable such laboratories to carry out their functions under sub-section (2) of section 12;

(g) the qualifications of Government Analyst appointed or recognised for the purpose of analysis of samples of air, water, soil or other substances under section 13;

(h) the manner in which notice of the offence and of the intention to make a complaint to the Central Government shall be given under clause (b) of section 19;

(i) the authority of officer to whom any reports, returns, statistics, accounts and other information shall be furnished under section 20;

(j) any other matter which is required to be, or may be, prescribed.

26. Rules Made Under This Act to be Laid Before Parliament

Every rule made under this Act shall be laid, as soon as may be after it is made, before each Hose of Parliament, while it is in session, for a total period of thirty days which may be comprised in one session or in two or more successive sessions, and if, before the expiry of the session immediately following the session or the successive sessions aforesaid, both Houses agree in making any modification in the rule or both Houses agree that the rule should not be made, the rule shall thereafter have effect only in such modified form or be of no effect, as the case may be; so, however, that any such modification or annulment shall be without prejudice to the validity of anything previously done under that rule.

Selected References

1. J.O.M.Bokris; Environmental Chemistry Plenum Press, New York, 1977
2. H.M.Dix; Environmental Pollution, John Wiley and Sons, New York, 1981
3. R.H.Perry and C.H.Chilton; Chemical Engineers Hand Book 5th and 6th Edition, McGraw Hill, New York, 1973, 1990
4. R.E.Trebal; Mass Transfer Operation, Second Edition, McGraw Hill, New York, 1952
5. M. Crawford; Air Pollution Control Theory, McGraw Hill, New York, 1976
6. E.D. Schroeder; Water and Wastewater Treatment McGraw Hill, New York, 1977
7. Octave Levenspiel; Chemical Reaction Engineering, 3rd ed., John Wiley and Sons, 1999
8. McCabe and Smith; Unit Operation of Chemical Engineering 6th ed., McGraw Hill, 2001
9. MetCalf and Eddy; Wastewater Engineering, Disposal and Reuse, Tata McGraw Hill, India, 1999
10. M.N.Rao and A.K.Tatta; Wastewater Treatment 2nd ed., Oxford and IBH Publishing Company, India, 1987
11. M.N. Rao; Air Pollution, McGraw Hill, 1993
12. W.L. Badger and J.T Banchero; Introduction to Chemical Engineering, McGraw Hill, 1993

Suggested References for Further Reading

1. MetCalf and Eddy; Wastewater Engineering, Disposal and Reuse, Tata McGraw Hill, India, 1999
2. C.S. Rao; Environmental Pollution Control Engineering, New Age International (P) Ltd, New Delhi, 1991
3. M.N. Rao; Air Pollution, McGraw Hill, 1993
4. G.M.Masters; Introduction to Environmental Engineering and Science Prentice Hall of India (P) Ltd, New Delhi, 2005
5. Gerald Kiely; Environmental Engineering, McGraw Hill, 1998
6. J.Glynn Henry and Gary W.Heinke; Environmental Science and Engineering, Pearson Education, 1996

Zeitfracht Medien GmbH
Ferdinand-Jühlke-Straße 7
99095 Erfurt, Deutschland
produktsicherheit@kolibri360.de